GPS
FOR
DUMMIES®
2ND EDITION

by Joel McNamara

Wiley Publishing, Inc.

GPS For Dummies,® **2nd Edition**

Published by
Wiley Publishing, Inc.
111 River Street
Hoboken, NJ 07030-5774

www.wiley.com

Copyright © 2008 by Wiley Publishing, Inc., Indianapolis, Indiana

Published by Wiley Publishing, Inc., Indianapolis, Indiana

Published simultaneously in Canada

For general information on our other products and services, please contact our Customer Care Department within the U.S. at 800-762-2974, outside the U.S. at 317-572-3993, or fax 317-572-4002.

For technical support, please visit www.wiley.com/techsupport.

Wiley also publishes its books in a variety of electronic formats. Some content that appears in print may not be available in electronic books.

Library of Congress Control Number: 2008936356

ISBN: 978-0-470-15623-0

Manufactured in the United States of America

10 9 8 7 6 5 4 3 2 1

WILEY

GPS

FOR

DUMMIES®

2ND EDITION

About the Author

Joel McNamara first got involved with digital maps in the early 1980s. At the time he was studying archeology and instead of going out and playing Indiana Jones, he found himself in front of a computer monitor trying to predict where archeological sites were located based on LANDSAT satellite data.

The lure of computers ultimately led to his defection from academia to the software industry, where he worked as a programmer, technical writer, and manager; eventually ending up at a rather large software company based in Redmond, Washington. Joel now writes and consults on technology he finds interesting, such as GPS and digital maps.

Over the years he's had practical experience using GPS and maps for wild-land firefighting, search and rescue, and disaster response and planning. He's also an avid user of the great outdoors (which means there's way too much gear in his garage). He likes boats of all kinds and is fond of certain old-school, dinosaur technologies such as celestial navigation.

Joel is also the author of *Geocaching For Dummies*, *Secrets of Computer Espionage: Tactics & Countermeasures*, and *Asus Eee PC For Dummies* (all published by Wiley).

Author's Acknowledgments

Once again, thanks to my wife, Darcy, for all her support during my work on this book; especially for the patience in putting up with all of the maps, CD-ROMs, cables, manuals, and gadgets that were scattered all over the house.

Next on the list are the folks at Wiley, including Katie Feltman, my acquisitions editor at Wiley, who was happy with the first edition of this book and asked me to update it. Mark Enochs, project editor extraordinaire, ably kept the book on course. And Dan Kearl, a savvy search and rescue/disaster response colleague, GPS instructor, and all-around techie who kept me honest as my technical editor. (A note of appreciation to Pat O'Brien, project editor, and GPS and map guru Gavin Hoban, technical editor, for all the work they did on the first edition.)

I'd also like to thank the following manufacturers for supplying review copies of their products to write about in the first and second editions: DeLorme (Caleb Mason and Charlie Conley), Endless Pursuit, Lowrance (Steve Wegrzyn), Magellan, Maptech, Microsoft, National Geographic, and TopoFusion (Scott Morris). I'm especially grateful for the help from the folks named in the parentheses who went above and beyond the call of duty in answering questions and providing assistance.

Finally, thanks to everyone who gave me feedback on various parts of this edition and the previous one. You know who you are, and I appreciate it.

Publisher's Acknowledgments

We're proud of this book; please send us your comments through our online registration form located at www.dummies.com/register/.

Some of the people who helped bring this book to market include the following:

Acquisitions and Editorial

Senior Project Editor: Mark Enochs

Senior Acquisitions Editor: Katie Feltman

Copy Editor: Brian Walls

Technical Editor: Dan Kearl

Editorial Manager: Leah Cameron

Editorial Assistant: Amanda Foxworth

Sr. Editorial Assistant: Cherie Case

Cartoons: Rich Tennant
(www.the5thwave.com)

Composition Services

Project Coordinator: Erin Smith

Layout and Graphics: Reuben W. Davis, Melissa K. Jester, Christin Swinford, Christine Williams

Proofreaders: Christine Sabooni, Amanda Steiner

Indexer: Claudia Bourbeau

Publishing and Editorial for Technology Dummies

    Richard Swadley, Vice President and Executive Group Publisher

    Andy Cummings, Vice President and Publisher

    Mary Bednarek, Executive Acquisitions Director

    Mary C. Corder, Editorial Director

Publishing for Consumer Dummies

    Diane Graves Steele, Vice President and Publisher

Composition Services

    Gerry Fahey, Vice President of Production Services

    Debbie Stailey, Director of Composition Services

Contents at a Glance

Table of Contents

Introduction

<p>· ·</p>

*Y*ou may have guessed from the title that this book is about GPS (the satellite-based Global Positioning System) and maps (digital maps to be exact).

I remember back in 1989 when Magellan introduced the first handheld GPS receiver, the NAV 1000. (Don't worry. This isn't going to be one of those "I used to walk 20 miles to school in the snow when I was your age," stories.) The NAV 1000 was the size of a brick, and weighed a little less than two pounds. It was a single channel receiver, could only track four satellites, and supported only latitude and longitude coordinates. It could save 100 waypoints, and you could have a single route with up to 10 waypoints. It cost $2,500.

Fast-forward to the present: Now I can go down to my neighborhood sporting goods or electronics store and buy a GPS receiver the size of a small cellphone. It weighs a couple of ounces, can track three to four times as many satellites, and on a good day can tell me exactly where I'm located to within about 10 feet; and in several different coordinate systems, by the way. It supports at least 500 waypoints and 20 routes, with 125 waypoints apiece. Best of all, it costs around $100. And, for not too much more, I can get an automotive GPS receiver that tells me aloud how to get where I'm going. It's like living in the future.

Maps have followed the same evolutionary path. Paper maps have turned digital, and now you can visit a Web site and print a map with driving directions to just about anywhere for free. For under $100, you can buy mapping software that has a collection of CD-ROMs with detailed topographic maps that fully cover any state in the United States. Aerial and satellite photographs are readily available over the Internet, and stunning three-dimensional maps can be created with a few mouse clicks. Once the exclusive domain of professional cartographers and GIS (Geographic Information System) specialists, the average computer user can create and use digital maps with relative ease. A number of free and inexpensive programs make *desktop mapping* a reality for the rest of us.

So does all this mean we're entering the dawn of a new era where no matter where you are it's going to be hard to get lost? Well, yes and no.

Over the past several years, GPS receivers have become extremely popular and affordable. Lots of people who venture away from urban areas are carrying them. Cars come installed with GPS navigation systems for negotiating city streets and highways. Many cellphones have tiny GPS chips embedded in them. And, even if you don't have a GPS receiver, you can always go out on the Web and print a map of where you want to go. But, there are a few hitches in this perfect, always-found world:

- GPS receivers often boast so many features it's easy to get lost trying to figure them all out. Plus, most GPS receiver owners typically use only a small subset of the available features (and sometimes don't even know how to use these features well enough to avoid getting lost).

- GPS receivers have capabilities and limitations that many owners (or potential owners) really don't understand. This leads to frustration or not being able to use the devices to their full potential.

- Digital street map data displayed and used by GPS receivers is often updated (especially in areas experiencing lots of growth).

- Although many people have a general knowledge of how to read a map, at least the simple road variety, most don't know how to really maximize using a map.

- And finally, the average computer user isn't aware of the wealth of easy-to-use, free or inexpensive mapping resources he could be using to stay found.

The purpose of this book is to help you better understand and use GPS receivers and open your eyes to the world of digital mapping — and, I hope, put you on the path of always staying found or finding what you're looking for.

Who This Book Is For

If you're browsing through this book at your favorite bookstore right now, and are pondering whether to take it to the cashier, ask yourself these questions:

- Are you considering purchasing a GPS receiver?

- Have you recently purchased a GPS receiver (or got one as a gift)?

- Have you owned a GPS receiver for a while, but want to get more from it?

- Are you interested in using digital maps for your profession or hobby?

If you answered yes to any of these questions, then stop reading and immediately proceed to the cash register because this book will make your life easier (if you're still not convinced, feel free to continue flipping through the pages to see what I mean).

Getting a bit more specific, people in the following groups should find this book especially useful:

- ✔ *Recreation* – Hikers, hunters, fishers, mountain bikers, trail runners, cross-country skiers, snowshoers, snowmobilers, prospectors, pilots, boaters, geocachers, and anyone else who ventures outdoors away from cities and streets (with or without a GPS receiver).

- ✔ *Transporation* – Drivers of all types of motorized vehicles (cars, trucks, motorcycles, you name it) are increasingly becoming dependent on GPS to help them navigate highways, byways, and even off-road.

- ✔ *Commercial* – Land developers and real estate agents who are interested in the competitive advantage maps can bring them for planning or marketing purposes.

- ✔ *Government* – Emergency response agencies (search and rescue, fire, law enforcement, disaster relief) and urban planners who use maps as part of their planning and response activities.

- ✔ *Environmental* – Conservation agencies, organizations, consultants, and scientists (biologists, botanists, and other *ists*) who use maps for resource management and research.

- ✔ *Technology* – Anyone who likes to play with cool technology.

You may have noticed I didn't mention people like surveyors or GIS professionals. If your job primarily focuses on GPS and/or maps, you'll probably discover a few things in the following pages, but just remember that this book is for the average computer user and GPS receiver owner who don't have your level of technical experience, proficiency, and skills. Please don't expect to find the nuts and bolts of using GIS software or precision surveying electronics.

And finally, if you purchased the first edition of *GPS For Dummies* (thank you very much), you'll find I've made a number of updates and added a lot of new information to stay current with the latest in GPS and digital mapping products and services.

Setting Some GPS Expectations

Before getting started, I want to set a few expectations about the content you'll be reading about that relates to GPS receivers, just so we're all on the same page:

- ✔ This book focuses on handheld, consumer GPS receivers (typically used for land navigation) and automotive navigation systems. In addition to these types of GPS receivers, larger and less portable models are used in airplanes, boats, and commercial vehicles. The U.S. government and military uses restricted-use GPS units, and expensive receivers are used for surveying. Although some of these GPS receivers are discussed briefly, don't expect to find out as much about them as the land and auto consumer models.

✔ Although most GPS receivers have the same general functionality, there are a lot of differences in manufacturer and model user interfaces. In a way, it's like sitting a computer novice in front of three PCs, one running Microsoft Windows, one running Linux (with the KDE or Gnome interface), and the other a Macintosh, and then asking the volunteer to perform an identical set of tasks on each computer. Good luck! Because of the differences, you're not going to find detailed instructions on how to use specific GPS receiver models. What you will find is information on how to use most any GPS receiver, with some kindly suggestions tossed in when it's appropriate to consult your user guide for details.

✔ Finally, don't expect me to tell you what the best GPS receiver is. Like any consumer electronics product, GPS receiver models are constantly changing and being updated. Instead of recommending that you buy a certain brand or model (that could possibly be replaced by something cheaper and better over the course of a few months), I tell you what questions to ask when selecting a GPS receiver and give you some hints on which features are best for different activities. You can apply these questions and selection criteria to pretty much any GPS receiver (no matter how much the marketplace changes) to pick the right model for you.

Take comfort in the fact that it's pretty hard to go wrong when you purchase a GPS receiver from one of the brand-name manufacturers who specialize in GPS (Garmin, Magellan, Lowrance, and TomTom to name a few of the bigger players). All these companies make excellent products, and you can expect to get a number of years use from them. (The good news is that GPS technology and product features haven't changed as rapidly as personal computers. I can go out and happily use a GPS receiver from 1998, whereas the same vintage personal computer would have been recycled long ago. It may not lock on to satellites as fast or have as many new whiz-bang features, but it still tells me where I'm at.)

How This Book Is Organized

This book is conveniently divided into several different parts. The content in each part tends to be related, but feel free to skip around and read about what interests you the most.

Part I: All About Digital Maps

This part of the book introduces you to digital maps; actually, it presents some important universal concepts that apply to both paper and digital maps, such as coordinate systems, datums, and how to read and use maps. The focus is primarily on land maps but there are a few brief mentions of nautical and aeronautical charts. In this part, you find out about different types of digital maps that are available, especially the free ones you can get from the Internet, and about some of the software you can use for digital mapping.

Part II: All About GPS

This is the part of the book devoted to demystifying GPS and GPS receivers. You find out about the technology behind GPS (including its capabilities and limitations), basic GPS concepts (such as waypoints, routes, tracks, and coordinate systems), and how to select and use a handheld and automotive GPS receiver. I tell you all about GPS cellphones, PDAs (like Pocket PCs and Palms), trackers, and other devices, and a bit on the popular GPS sport of geocaching.

Part III: Digital Mapping on Your Computer

In this part, I take some of the theoretical information on digital maps from Part I, and get practical. This section discusses computer requirements needed for basic digital mapping and reviews a number of different software packages you can use to work with aerial photos, topographic maps, and road maps. Many of these programs support uploading and downloading data to and from GPS receivers, so I also spend some time talking about how to interface a GPS receiver to a personal computer.

Part IV: Using Web-Hosted Mapping Services

Even if you don't have a GPS receiver or mapping software installed on your computer, with an Internet connection and a Web browser you can still do a remarkable amount of digital mapping with free and subscription Web services. This section discusses how to access and use online street maps, topographic maps, aerial photos, and some slick U.S. government-produced maps. You also discover how to save and edit these Web-based maps.

Part V: The Part of Tens

All *For Dummies* books have a part called The Part of Tens, and this one is no exception. In this section, you find lists of what I consider the best GPS and digital map Web sites on the Internet, where to find free digital maps, and tips and hints on printing maps. And, if you're a boater, there's a lockerfull of marine GPS resources.

Icons Used in This Book

Maps use symbols to convey information quickly, and this book does the same by using icons to help you navigate your way.

Just a gentle little reminder about something of importance, and because I can't be there to mention it in person and give you a friendly (and stern if needed) look while wagging my finger, this icon will have to do.

I try to keep the real geeky, nerdy things to a bare minimum, but because this is a book about cool electronic gadgets and computer mapping, sometimes the technical stuff does creep in. I either give you a plain-English explanation or point you to a Web site where you can get additional details.

This is good stuff designed to make your life easier; usually gained from practical experience and typically not found in manufacturer user guides and product documentation; or if it is there, it's buried in some obscure paragraph.

The little bomb icon looks as if it should signify some pretty bad juju, but in reality it could represent something as minor as *potentially causes a hangnail*. The key here is to pay attention because there might be something lurking that causes mental, physical, emotional, or monetary suffering of some degree. Who would have thought reading *GPS For Dummies* could be an extreme sport?

Some Opening Thoughts

Before you jump into the exciting world of GPS and digital maps (I know you can't wait), I want to mention a few final thoughts:

- ✔ There are lots of references to Web sites in this book. Unfortunately, Web sites change just about as fast as street maps in a city experiencing a lot of growth. If for some reason a link doesn't work, you should have enough information to find what you're looking for by using common sense and a search engine, such as Google.

✔ You're not going to find every GPS and map software title in existence mentioned in the book. I try to list and describe many of the more popular programs, but the realities of page count constraints prevents this book from turning into an encyclopedia. So please don't get upset if I don't mention a program you use or you feel slighted because I end up talking about one program more than another.

✔ On some occasions while you're reading this book, you're probably going to think I sound like a broken record on one point I feel is very important. If you venture away from civilization with your GPS receiver, please bring a compass and a paper map with you, and know how to use them. That means really knowing how to use them, not just kidding yourself that you do. From many years of doing search and rescue work and finding lost people, I've discovered the following truths:

- GPS receiver batteries die at the most inopportune time; especially when you didn't bring spare batteries with you.

- If a GPS receiver breaks or gets lost, it will do so at the worst possible moment.

- GPS receivers are not *Star Trek* teleporters that will instantly transport you out of the wilderness and trouble (this is also true when it comes to cellphones).

✔ All the information in this book should set you on your way to becoming an expert with a GPS receiver and maps. That is, if you practice! If you want to have guru status, you need to be applying what you discover in this book. Even if you don't aspire to becoming one with GPS and a master of maps, to get the most use from your navigation tools, you need to become both comfortable and confident with them. Discover, experiment, and have fun!

Part I

All About Digital Maps

The 5th Wave   By Rich Tennant

"Of course your current GPS receiver downloads coordinates, prints maps, and has a built-in compass. But does it shoot silly string?"

In this part . . .

Although digital maps are made of bits and bytes, they share a number of things in common with their paper and ink cousins — like datums, coordinates systems, scales, legends, and compass roses. In fact, when you get some of these concepts down, you'll be at home with just about any map you encounter, whether it's displayed on your PC's monitor or laying on the front seat of your car.

Paper maps have a certain old school charm, but digital maps are infinitely cooler. That's because you can associate data with a digital map and make it interactive and smart. This part sets the stage for other chapters in the book. We're going to be talking about all sorts of PC, Web-based, and GPS maps, and it's important that you understand the basics of how digital maps work and what types of digital maps are out there; especially the free ones available on the Internet.

Chapter 1

Getting Started with Digital Maps

*T*his chapter introduces you to the fundamentals of digital maps. You find out what a digital map is, what the differences are between static and smart digital maps, and what different types of programs are available for using digital maps.

Amerigo Vespucci Didn't Have Maps like This

Any *map* is a picture of where things are, generally associated with our planet and its geographic or man-made features. Road maps, hiking maps, maps to homes of Hollywood stars, and all sorts of other maps provide a sense of place and often help you get from one location to another.

Traditionally, maps have been printed on paper, which makes them pretty convenient. They can be folded into a lightweight, compact bundle (if you've had a little practice). *Digital maps* (maps made on a computer or meant to be used with a computer) serve the same purpose as their paper cousins. It's just more difficult to fold a CD.

21st century cartography

Cartography is the art and science of making maps. Until the 1960s, maps were made the time-honored, traditional way:

1. Draw an original map by hand, based on land survey measurements and other information.

2. Print as many copies as you need.

That approach changed with the advent of computers, satellite imagery, and Global Positioning System (GPS), which made map-making much easier. Most paper maps now are generated or produced on a computer.

Digital mapmaking is a significant leap forward from traditional paper maps.

✔ **Maps can be made faster, cheaper, and more accurately.**

This is important because of how quickly new roads, subdivisions, and development projects pop up in fast-growing urban areas. An old street map isn't much help in a new subdivision with a couple hundred homes. The same problem affects political maps; an example is the change in national names and borders after the fall of the Soviet Union.

✔ **Digital map data can be used with mapping software to make digital maps on your personal computer.**

Static map

A *static map* is the simplest form of digital map — a paper map that's been scanned and turned into a BMP (bitmap) or JPG (graphic) file. Or, a static map can be a digital version of an aerial or satellite photo. Aside from displaying it, printing it, and perhaps making a few edits, what you can do with a static map is limited.

Nevertheless, a static map is often all you need.

Smart map

Smart digital maps (as shown in Figure 1-1) may look like static maps, but they have data associated with locations on the map. The data can be as basic as the latitude and longitude of a single point, or as detailed as providing information about vegetation, soil type, and slope.

Spatial or *geospatial* data is associated with a place. The place can be smaller than a meter (that's about 3 feet for the metrically challenged) or as large as a country. Spatial data can be stored two ways:

✔ Embedded in a map graphic file

✔ Stored in separate files with references to the locations

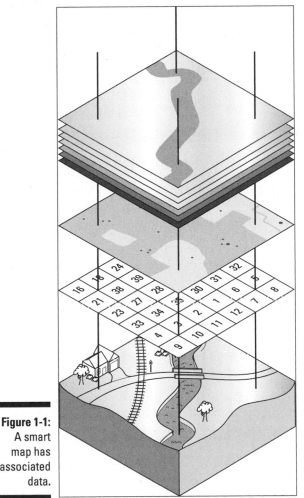

Figure 1-1:
A smart
map has
associated
data.

TIFF (Tagged Image File Format) is a popular format for storing graphics files. The *GeoTIFF* extension embeds geographic tags into map images. If you view a GeoTIFF file with a standard graphics program, it looks like an ordinary map. However, a program that uses the data tags can access the spatial data (such as latitude/longitude, elevation, and so on) associated with each pixel in the image.

Mapping Programs

Although many different kinds of mapping programs are available, you can classify map programs in two types: consumer programs and Geographic Information System (GIS) software. Here's a quick look at each type.

Consumer programs

A *consumer* mapping program is software that displays street maps, topographic maps, aerial/satellite photos, marine charts, or aeronautical charts. Such mapping programs are easier to use (and much less expensive) than their professional counterparts, meeting most computer users' mapping needs. This book focuses on mapping programs available to consumers.

GIS (Geographic Information System) programs

A *Geographic Information System* (GIS) is an information system that analyzes, inputs, manipulates, outputs, retrieves, and stores spatial data. GIS is mostly used by governments; large corporations; and engineering and GIS consulting firms for land, natural resources, transportation, and urban planning.

Some people use the terms *digital map* and *GIS* interchangeably. This really isn't correct. GIS isn't just about making maps. GIS involves using computers and special software to help people make decisions by using spatial data.

Distinguishing between consumer mapping programs and GIS programs is important:

 ✔ **Consumer mapping programs** target the needs of average computer users. These programs are much more limited in scope and functionality — and a lot less expensive — than GIS programs.

 ✔ **GIS software,** which is sold primarily to governments, corporations, and consulting firms, is flexible, powerful, and relatively expensive.

 GIS software typically has a steep learning curve; you can earn advanced degrees in GIS. Consumer mapping programs can mostly be used right out of the box and can be mastered in a relatively short period.

Consumer versus GIS programs

A typical consumer mapping program is a road map program that costs about $30 and provides exact routing directions to get from one location to another. This isn't a static map because it has underlying data (such as street names, distances, and gas stations), which can lead you to think it's a GIS program. Not so. A true GIS program has built-in precision tools that can (for example) let you input data about traffic flow and vehicle speeds, and then display every street where traffic volume exceeds 500 cars per hour and vehicle speeds are

.5 miles an hour over the speed limit. The price tag for such a GIS program would be at least $1,000, not to mention the costs of training people to use it and gathering all the traffic data to input into the system.

Of course, if you have a burning need for high-end precision and complexity, it's still possible to get into GIS on the cheap. A growing community is developing open source and free GIS programs. Although many of these programs lack the polish of a commercial product, they do get the job done. The http://opensourcegis.org and www.freegis.org Web sites are two excellent resources for finding out more about free GIS programs.

Digital Maps in Practice

There's an old song that goes, "Anything you can do, I can do better." If digital maps could sing that tune to their paper counterparts, they'd be right (for the most part). Digital mapping software offers all sorts of enhancements over paper maps, including these capabilities:

- ✔ Finding street addresses quickly
- ✔ Interfacing with GPS receivers to see where you are or where you were
- ✔ Showing driving directions to just about anywhere
- ✔ Displaying terrain three-dimensionally
- ✔ Annotating maps with pop-up information (spatial data)
- ✔ Creating custom maps
- ✔ Printing a hard copy map

Digital maps do have a few drawbacks, including these:

- ✔ **You need a computer.**

  If you have a laptop or personal digital assistant (PDA), or map-enabled cellphone, you can take mapping software on the road with you. Just remember you need power to use these maps. So make sure your batteries are fully charged before heading out.

- ✔ **You need software.**

  This book helps you select and use software packages, particularly mapping programs in the free–$100 range.

- ✔ **You have to spend time mastering the software.**

  Most mapping software is readily usable, but all programs have nuances that sometimes make their features and user interfaces a little tricky.

Mapping Software: The Essentials

The first step for digital mapping is to understand the available types of mapping programs and their capabilities and limitations; that's what this part of the book is all about.

After you know what software is available, you can match it to your needs. An invitation to a birthday party might consist only of displaying a screen capture of a street map on a Web site, editing and saving the map in Paint, and then e-mailing it to friends. A weeklong backpacking expedition would require a *topographic* mapping program (showing land features) to plan your route, view elevation profiles, and upload location data to your GPS receiver.

Before you can select the right tool for the right job, you need a general handle on the options that you can include in your digital-mapping tool chest. This section of the book organizes mapping programs into three categories:

- ✔ Standalone programs
- ✔ Programs bundled with maps
- ✔ Web-hosted mapping services

Standalone programs

A *standalone program* is a program that can open and use digital maps. These programs typically don't come with map data; therefore, you need to download or purchase the maps you're interested in using.

Like a word processor or a spreadsheet, a mapping program needs someone to input data before it can be useful. In this case, the data is bits and bytes that describe how a map should display. Fortunately, an amazing amount of map data is freely available on the Internet, most of it already collected by the government and in the public domain.

A big market exists for commercial map data. People buy data to use with their mapping programs because

- ✔ Free data may not be available for an area or a specific need.
- ✔ Commercial data might be enhanced with information unavailable in the free versions. Additionally, companies offering commercial map data often release updated versions of their data — and let you know when it's available.
- ✔ Firing up a CD filled with data is more convenient and faster than searching for, and then downloading, free data.

Many standalone mapping programs aren't tied to one data type. (Chapter 2 shows which types of digital map data are commonly used.) Figure 1-2 shows a three-dimensional map of Mount St. Helens created with a program called 3DEM from free U.S. Geological Survey (USGS) digital elevation map (DEM) data. The elevation map shows the crater and the blown-out side from the volcano's 1980 eruption.

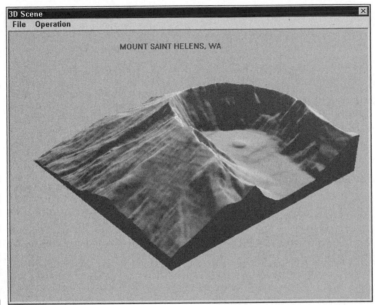

Figure 1-2:
A 3-D map
created with
a free map
program
and free
map data.

Map programs are viewers, editors, or both:

- ✔ **Viewers** only show maps.

- ✔ **Editors** allow you to make changes.

    Usually, you can't change a base map you've opened from a data file, but you can add text and draw shapes on the map.

Many standalone programs are either free or shareware. Two aspects of such programs are especially worth noting:

- ✔ Some manufacturers offer free (or cheap) limited-feature versions of their products that are otherwise available as pricy commercial software. These programs are just as good as the commercial versions for basic purposes.

- ✔ Standalone mapping programs are mostly suited to a user who has intermediate to advanced computer skills and experience. Make sure you get one that suits your skills.

Examples of standalone mapping programs include OziExplorer (www. oziexplorer.com), USAPhotoMaps (http://jdmcox.com), and 3DEM (www.visualizationsoftware.com/3dem.html), which made the spiffy image in Figure 1-2. Don't forget that you can also make maps with Paint or any other general-purpose graphics program. This book talks about these programs and others.

Some free, noncommercial mapping programs have advanced features that are normally more suited to professional users. Don't be intimidated by lots of features and options. You can use a few commands and features to make maps that meet your needs. Master the other features if you ever need to.

One requirement for working with standalone mapping programs is that you need to search the Internet for suitable data, find and download it, and then open it with the map program. This process sometimes involves *registering* or *georeferencing* a map so that the coordinates are accurate. And even with a high-speed Internet connection, downloading can take a while. After you have the data downloaded, you still have to find the map data for an area that you want to view, and then successfully load all that stuff into the mapping program.

Programs with bundled maps

Mapping companies bundle software with digital maps. The program comes with the map data and is distributed on CDs or DVDs; static or smart maps that have a lot of detail can be quite large. You install the mapping program, and you're immediately ready to start using the data on the CD.

Data files bundled with software are often in a proprietary file format, which can be read and used only with the software that comes with the product. The same usually holds true for maps that you can upload to a particular brand of GPS receiver; only maps from the manufacturer can be used.

Sometimes you don't have much choice between using a standalone program or one bundled with maps. However, keep the following in mind if cost is an issue:

- Free topographic map data of the United States is widely available. You can use a number of free or shareware programs to view maps.

- Typically, outdated Census Bureau map data is available for United States streets and roads. Most free or shareware programs don't match the features in commercial street map products.

Software that comes with bundled maps has gotten incredibly cheap over the years. With discounts and rebates, you can often find road atlas software for around $20 that covers the entire United States. For a little under $100, you can buy programs that come with a full set of detailed digital topographic maps for an entire state (or even larger region). Considering a single paper USGS 1:24,000 map costs around $7 — and there can easily be over a thousand maps per state — that's a pretty decent value. Figure 1-3 shows a map made by a program called Terrain Navigator (www.maptech.com/land/ TerrainNavigator), which is a topographic mapping program that comes bundled with map data.

Figure 1-3:
A 1:100,000 scale topographic map displayed with Maptech's Terrain Navigator.

Manufacturers that sell bundled map programs (particularly those with street and road data) usually come out with a new release of their product every year or so. In addition to enhancements in the software, the map data contains new roads and updated services information, such as gas stations, restaurants, and hotels (called *POIs*, or *Points of Interest*). Whether you buy an updated copy of the software every year depends on your circumstances. If you usually travel on major roads, or in areas that haven't experienced much development and growth, you probably don't need to update every year. On the other hand, road atlas software is fairly inexpensive, so if you travel a lot and rely on the program, it can be a cheap investment.

If you have beginning to intermediate computer skills and experience, you can come up to speed quickly with bundled map programs. The user interfaces are generally simpler than those found in feature-rich, standalone programs.

Examples of programs that come bundled with maps are DeLorme's Street Atlas USA (www.delorme.com), National Geographic's TOPO! (www.natgeomaps.com/topo), and mapping software from GPS manufacturers that interfaces with their receivers.

In addition to these programs, applications that stream data off the Internet can display maps. The map data doesn't physically come with the program, but is virtually bundled. Examples include Google Earth (`earth.google.com`) and TopoFusion (`www.topofusion.com`).

I discuss both types of programs in Part III.

Web-hosted mapping services

A *Web-hosted mapping service* is a Web site that displays a map. You just need Internet access and a browser to view street maps, topographic maps, aerial maps, satellite imagery, and many other types of maps. This eliminates purchasing and installing specialized programs and map data on your hard drive, swapping CDs to access new map data, and mastering a new program. (Figure 1-4 is a detailed street map of downtown Port Angeles, Washington, using `maps.google.com`, a Web-hosted mapping service.)

If a map isn't displayed, check your browser's Java settings first. A number of Web-hosted mapping services, in particular the U.S. government sites that all share the same mapping engine, require Java and/or JavaScript enabled in your browser before maps can be correctly displayed.

A few words on commercial GIS software

This book doesn't dwell overmuch on commercial GIS software packages. If your mapping needs get complex enough to require GIS (or you just want to find out more about these high-end mapping systems), check out the Web sites of the top three GIS companies:

✔ **ESRI:** Environmental Systems Research Institute is the largest GIS company in the marketplace. Its Arc products (such as ArcInfo and ArcView) are standards in the GIS field. For more information about ESRI, go to `www.esri.com`.

✔ **MapInfo:** MapInfo develops and sells a wide array of GIS products. For more information on the company and its products, see `www.mapinfo.com`.

✔ **Autodesk:** Autodesk is the developer of AutoCAD, a widely used computer-aided design program. Autodesk add-ons and standalone GIS programs are used throughout the world. You can find out about the Autodesk mapping applications at `www.autodesk.com`.

The GIS Lounge is a great Web resource that isn't tied to a specific manufacturer. This site provides information on all aspects of GIS, and has informative and educational content for novices to professionals. For more information, go to `www.gislounge.com`.

Or, if you prefer books, be sure to check out *GIS For Dummies* by Mike DeMers (Wiley Publishing).

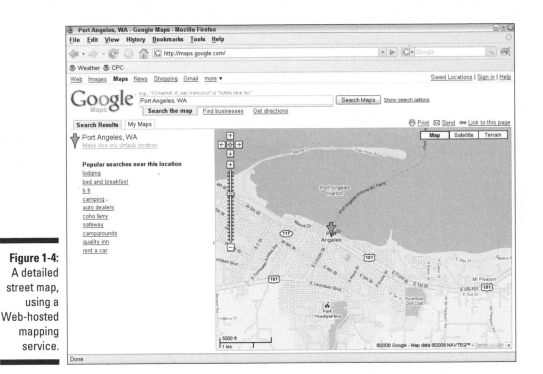

Figure 1-4:
A detailed
street map,
using a
Web-hosted
mapping
service.

Most Web-hosted mapping services are extremely easy to use. Anyone who can use an Internet browser should be navigating through maps in no time.

Examples of Web-hosted mapping services include Google Maps (maps. google.com), TerraServer-USA (http://terraserver-usa.com), and TopoZone (www.trails.com). You discover how to use these mapping Web sites and others in Part IV.

Although most of these free Web sites don't have all the features of a map program that you install on your hard drive, they offer a surprising amount of capability, especially considering their cost. (Some mapping sites on the Web charge for advanced services, such as color aerial photographs, larger map sizes, and enhanced searching.)

Chapter 2

Dissecting Maps

. . . .

. . . .

Maps are everywhere. They tell us how to get to places we want to go and give us a better understanding of our surroundings.

Some people think they can pick up a map and immediately start using it. If you don't have a lot of experience using different types of maps, that's a little like thinking just because you can read, you can sit down and understand a book that's written in French, German, or Spanish even if you've never had a foreign language class in your life. Yes, you might make a little sense of the book by picking out a few words you recognize or by looking at the pictures, but in the process, you're missing out on some important information.

That's what this chapter covers: the different types of maps, basic map concepts and principles, and the various kinds of digital maps that you can access on a computer. You'll have a handy enough grasp of conversational map-speak that you can ask the right questions to avoid getting lost.

Discovering Types of Maps

Maps have their own language and a number of dialects, depending on the type of map. To use paper or digital maps effectively, you need to have at least a tourist's understanding of their language. The more you know, the better off you are.

Begin by looking at the basic types of maps that you use to navigate with. (Although there are maps for traveling under the ocean, visiting the moon, and zooming around in space, you're not likely to need these anytime soon.)

An important point to consider is that no one universal map type does it all. Different map types display different features and details that are suited for a particular use — or user. A skilled map user always selects a map that meets his or her specific needs.

Maps are almost always oriented so the top of the map is facing north. If a map doesn't follow this convention, a good mapmaker places an arrow on the map that points north.

Land

Because we spend most of our time with our feet or tires on the ground, knowing how to use land maps is pretty important. Generally, the two types of land maps are topographic and planimetric.

Topographic maps

Often called *topo maps* for short, *topographic maps* show natural land features such as lakes, rivers, and mountain peaks as well as man-made features such as roads, railroad tracks, and canals. These maps also have contour lines that trace the outline of the terrain and show elevation. *Contour lines* suggest what the land looks like in three dimensions.

A *contour interval* is the distance between two contour lines. For example, if a contour interval is 20 feet, every time you go up one contour line, the elevation increases by 20 feet. Conversely, every time you go down a contour line, the elevation decreases by 20 feet. When the contour lines are close together, the terrain is steep. When they're spread apart, the terrain is closer to flat. Different maps have different contour intervals, and the distance is usually noted in the map legend.

The most popular topographic maps for use within the United States are made by the United States Geological Survey (USGS). These maps cover different sizes of area — the smaller the area, the greater the detail. The topo maps that show the most detail are sometimes called *quad sheets* or *7.5 minute maps* because they map just one *quadrangle* (geographer-speak for a rectangular-shaped piece of land) that covers 7.5 minutes of longitude and latitude. Figure 2-1, for example, is a portion of a topographic map of The Dalles, Oregon.

A compass uses degrees to tell direction. North is 0 or 360 degrees, west is 90 degrees, south is 180 degrees, and east is 270 degrees.

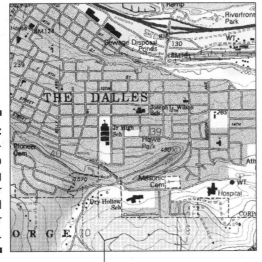

Figure 2-1:
A topographic map showing contour lines and other features.

Contour line

Most topographic maps show *magnetic declination* — in marine and aeronautical use, the term used is *variation*. Compass needles point to magnetic north, but most maps are oriented to true north. Because the Earth's magnetic field varies from place to place, magnetic north usually isn't the same as true north; in the continental United States, the difference can be as much as plus or minus 20 degrees. If you don't account for the magnetic declination, you can wander way off-course navigating with a compass. For example, when a location has a declination of 20 degrees and you don't account for it, your compass heading will be off by 20 degrees — over a distance, you may end up miles away from your planned destination. The declination tells you how many degrees you need to adjust your compass: If the declination is west, you subtract the degrees from 360 to get true north; if it's east, you add the degrees.

Magnetic declination changes over time, and older USGS maps can have incorrect declination information printed on them. Using the wrong declination can cause all sorts of navigation problems, so check the current declination for your area at the following Web site: www.ngdc.noaa.gov/seg/geomag/declination.shtml.

If your job or hobby takes you off the beaten path, you definitely need a topographic map. If you're staying in your car, driving on paved roads, you'll probably be fine with a planimetric map.

Planimetric maps

These are the maps you're most familiar with — you've just never heard the word *planimetric*. Planimetric maps don't provide much information about the terrain. Lakes, rivers, and mountain pass elevations may be shown, but there isn't any detailed land information. A classic example of a planimetric map is a state highway map or a road atlas. Planimetric maps are perfect in cities or on highways, but they're not suited for backcountry use. Figure 2-2 is a planimetric map of The Dalles, Oregon, area.

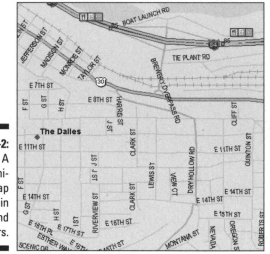

Figure 2-2: A plani-metric map lacks terrain features and contours.

When using planimetric maps, you often encounter these terms:

- **Atlas:** An *atlas* is a collection of maps, usually in a book.
- **Gazetteer:** A *gazetteer* is a geographical dictionary or a book that gives the names and descriptions of places.

Marine

Marine charts are maps for inland, coastal, and deep-water navigation. Charts from the National Oceanic and Atmospheric Administration (NOAA) are commonly used for boating. They provide such important information as water depth, buoy locations, channel markers, and shipping lanes. Part of a marine chart of San Francisco Bay, California, is shown in Figure 2-3. See `http://nauticalcharts.noaa.gov/` for more on NOAA charts.

Marine charts aren't available for all bodies of water. If you're boating on a small lake or a river, you'll probably use a topographic map for navigation.

If you're more of a sailor than a landlubber, Chapter 22 presents a number of GPS-related resources for mariners.

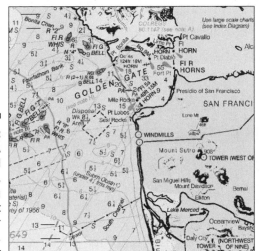

Figure 2-3:
A NOAA
nautical
chart,
showing
important
features for
mariners.

Aeronautical

Maps designed for aviation use are *charts* (a term that also refers to their marine counterparts). These maps provide pilots with navigation information including topographic features, major roads, railroads, cities, airports, visual and radio aids to navigation, and other flight-related data. Figure 2-4 shows a pilot's-eye view of the Seattle area.

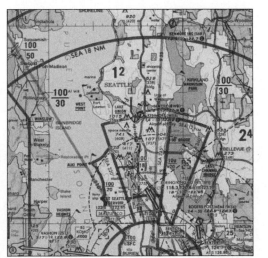

Figure 2-4:
A sectional
aeronautical
chart for
the Seattle
area.

Anatomy of a map

Most maps have some elements in common. Here are the ones you're most likely to run across, along with the terms that *cartographers* (mapmakers) use to describe them:

Citation: This is information about data sources used in making the map and when the map was made.

Collar: The white space that surrounds the neatline (see the upcoming definition) and the mapped area.

Compass rose: A map has either a simple arrow that shows north or a full *compass rose* (an image that indicates all four directions) so the user can correctly orient the map to a compass.

Coordinates: Maps usually have letters and numbers or coordinates (such as latitude and longitude values) marked along the borders so that users can locate positions on the map.

Legend: A box that shows an explanation of symbols used on the map. Some maps show all the map symbols; others rely on a separate symbol guide.

Mapped area: The main part of the map, displaying the geographic area.

Neatline: The line that surrounds the mapped area.

Scale: Distance-equivalence information (such as "one inch = one mile") typically found at the bottom of a map that helps you estimate distances.

Title: Usually the name of the map, but it also often tells you which area it's mapping.

Standard aeronautical chart types include

- ✔ **VFR (Visual Flight Rules):** Charts used for flying without instruments
- ✔ **IFR (Instrument Flight Rules) Enroute:** Charts used for instrument flying
- ✔ **Terminal Area Charts:** Charts covering areas around large airports

You can find more about aeronautical charts by visiting the Federal Aviation Administration (FAA) National Aeronautical Charting Office (NACO) at www. naco.faa.gov.

The FAA doesn't make aviation charts freely available, but you can download sectional charts in GeoTIFF format from the NACO Web site previously mentioned for $1.50 each. Jeppesen (www.jeppesen.com) and other companies make commercial flight-planning software packages that include digital charts, or you can try www.aeroplanner.com, a Web service that provides digital charts and other services to pilots. For free Web-based sectional charts, check out http://skyvector.com. Another free source of older FAA sectional charts is http://aviationtoolbox.org/raw_data/FAA_sectionals.

Figuring Out Map Projections

Making a map is quite a bit more challenging than you may think. A cartographer's first challenge is taking something that's round like the Earth (technically, it's an ellipsoid that bulges in the middle and is flat at the top and bottom) and transforming it into something that's flat, like a map.

Cartographers use a *projection* to reproduce all or part of a round body on a flat sheet. This is impossible without some distortion, so the mapmaker decides which characteristic (area, direction, distance, scale, or shape) will be shown accurately and which will be distorted.

Although my high school geography teacher may smack me on the head with a globe for saying this, the average map user doesn't need to know what kind of projection was used to make a map. There are some exceptions if you're a cartographer or surveyor, but usually you won't get in trouble if you don't know the projection. So don't panic if you can't immediately tell a Lambert conformal from a Mercator or Miller projection. Just keep in mind what a projection is and that there are different types of map projections.

Map Datums

A *map datum* is a mathematical model that describes the shape of an ellipsoid — in this case, the Earth. Because the shape of the Earth isn't uniform, there are over 100 datums for different parts of the Earth, based on different measurements.

Some serious math is involved for getting into the nuts and bolts of map datums. If you're the scholarly type, these Web sites provide lots of details on projections and datums:

- Datums

    ```
    http://geology.er.usgs.gov/eespteam/GISLab/Cyprus/
            datums.htm
    ```

- Peter Dana's excellent Geographer's Craft site

    ```
    www.colorado.edu/geography/gcraft/notes/notes.html
    ```

Datums all have names, and they often aren't as stuffy sounding as you might expect. In fact, some are rather exotic, with Indiana Jones–style names, such as the Kerguelen Island, Djakarta, Hu-Tzu-Shan, or Qornoq datums. (The United States uses the fairly mundane NAD 27 and WGS 84 datums.)

Concerning datums, there are three important points to remember:

 ✔ A location you're interested in is plotted on two different maps.

 ✔ A map and a Global Positioning System (GPS) receiver are being used.

 ✔ Two different GPS receivers are being used.

In these instances, all the maps and GPS receivers must use the same datum. If the datums are different, the location ends up in two different physical places even though the map coordinates are exactly the same.

Mixing datums is a common mistake you should avoid making: GPS receivers use the WGS 84 datum by default, and most older USGS topographic maps use the NAD 27 datum. If you mix the datums, your location can be off by up to 200 meters (roughly 200 yards, if you missed the lesson on the metric system back in school).

It is possible to make a mixed datum approach work. Utilities can convert coordinates from one datum to another (some are described in Chapter 12), but it's easier just to get all the datums on the same map, so to speak.

Working with Map Coordinate Systems

A *coordinate system* is a way to locate places on a map, usually by laying some type of grid over the map. Grid systems are a whole lot easier to use and more accurate than "take the old dirt road by the oak tree for two miles, then turn left at the rusted tractor, and you'll be there when the road stops getting bumpy."

A simple coordinate system can consist of a vertical row of letters (A, B, C) on the left side of the map and a horizontal row of numbers (1, 2, 3) at the bottom of the map. For example, if you want to tell someone where the town of Biggs Junction is, you put your finger on the city and then move it in a straight line to the left until you hit the row of letters. Then put your finger on the city again, but this time move down until you reach the row of numbers. You now can say confidently that Biggs Junction is located at A12.

I call this the *Battleship Grid System* because it reminds me of the game where you call out coordinates to find your opponent's hidden aircraft carriers, submarines, and destroyers. "B-3. You sank my battleship!"

A grid may be printed on the map or provide *tick marks* (representing the grid boundaries) at the map's margins. Often, maps have multiple coordinate systems so pick one that meets your needs or that you're comfortable using. For example, USGS topographic maps have latitude and longitude, Universal Transverse Mercator (UTM), and Township and Range marks.

Most coordinate systems are based on x and y; where *x* is a horizontal value and *y* is a vertical value. A location's coordinates are expressed by drawing a straight line down to x and across to y. (Mathematician René Descartes devised this system in the 1600s.)

Letter-and-number coordinate systems are fine for highway maps, road atlases, and other simple maps where precise locations aren't needed. However, if you want to focus on a pinpoint map location, you need a more sophisticated grid system — the sport of geocaching is an example. That's where coordinate systems, such as latitude and longitude and UTM, come in.

When you're figuring out a location's coordinates on a paper map, you have a fair amount of work to do, aligning the location with primary tick marks and then adding and subtracting to get the exact coordinate. With digital maps on a computer, all that manual labor is usually accomplished by moving the cursor over a location and watching with relief as the coordinates automatically appear. If you're using a paper map, you can make life easier with free overlay grids and rulers from www.maptools.com. With these, you can print grids and rulers for different coordinate systems on clear transparency sheets.

Latitude/longitude

Latitude and longitude is the oldest map-coordinate system for plotting locations on the Earth. The Roman scholar Ptolemy devised it almost 2,000 years ago. Seeking a way to accurately represent the Earth on a flat piece of paper, Ptolemy created latitude and longitude. That's pretty impressive for a time way before computers and satellites.

Latitude and longitude are based on a little math, but they're not really complicated. Angles are measured in degrees, and they're used for measuring circles and spheres. Spheres can be divided into 360 degrees; because the Earth is basically a sphere, it can also be measured in degrees. This is the basis of latitude and longitude, which use imaginary degree lines to divide the surface of the Earth (see Figure 2-5).

The *equator* is an imaginary circle around the Earth. This special circle is located an equal distance from the north and south poles and perpendicular to the Earth's axis of rotation. The equator divides the Earth into the *Northern Hemisphere* (everything north of the equator) and the *Southern Hemisphere* (everything south of the equator).

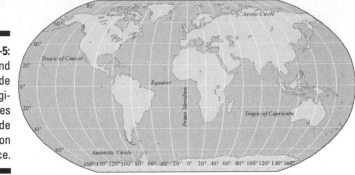

Latitude

Latitude is the angular distance measured north and south of the equator (which represents 0 degrees of latitude).

- As you go north from the equator, the *north latitude* increases to 90 degrees when you arrive at the North Pole.

- As you go south of the equator, the *south latitude* increases to 90 degrees at the South Pole.

In the Northern Hemisphere, the latitude is always given in *degrees north;* in the southern hemisphere, it's given in *degrees south*.

Longitude

Longitude works the same way as latitude, but the angular distances are measured east and west of the prime meridian (which marks the 0-degrees longitude line that passes through Greenwich, England, without even disturbing traffic).

- When you travel east from the prime meridian, the longitude increases to 180 degrees.

- As you go west from the prime meridian, longitude also increases to 180 degrees. (The place where the two 180-degree longitudes meet is the *International Date Line.*)

- In the Eastern Hemisphere (which is east of the prime meridian to 180 degrees east), the longitude is given in degrees east.

- In the Western Hemisphere (which is west of the prime meridian to 180 degrees west), longitude is expressed in degrees west.

By degrees, minutes, and seconds

One degree is actually a pretty big unit of measure. At the equator, one degree of latitude or longitude is roughly equal to 70 statute miles, or 60 nautical miles.

Too much latitude

Latitude and longitude are pretty straightforward and logical if you think about it. Unfortunately, over the years, people have muddied things a bit by inventing different ways to represent latitude and longitude coordinates.

Latitude and longitude coordinates can be written as

✔ **Degrees, minutes, and seconds:** This is the traditional way, with my example of Dillon Falls expressed as 43° 57′ 29.79″ N 121° 24′ 34.73″ W.

✔ **Degrees and decimal minutes:** Seconds are dropped, and the decimal version of minutes is used along with degrees, so now the falls are at 43° 57′ 29.79″ N 121° 24′ 34.73″ W.

✔ **Decimal degrees:** Minutes and seconds are both dropped, and only the decimal representation of degrees is used, which puts the falls at 43.9582750° N 121.4096490° W.

It's a good idea to always include an N or S following latitude and an E or W following longitude.

Remember: Although they look different, these coordinate notations point to the same location. The math is pretty straightforward to convert the coordinates from one format to another. If you want to save time, point your Web browser to `http://nris.state.mt.us/wis/location/latlong.asp` where the friendly people at the Montana State Library have a handy online conversion calculator.

Degrees are composed of smaller, fractional amounts that sound like you're telling time.

✔ **Degree:** A *degree* comprises 60 minutes.

One *minute* is about 1.2 miles.

✔ **Minute:** A *minute* is composed of 60 seconds.

One *second* is around .02 miles.

These distances are measured at the equator, in statute miles. At higher latitudes, the distance between longitude units decreases. The distance between latitude degrees is the same everywhere.

These measurement units are abbreviated with the following symbols:

✔ **Degree:** °

✔ **Minute:** ′

✔ **Second:** ″

If you use minutes and seconds in conjunction with degrees, you can describe a very accurate location. For example, if you're using latitude and longitude to locate Dillon Falls on a map of the Deschutes River in Oregon, its coordinates are

43° 57' 29.79" N 121° 24' 34.73" W

That means that Dillon Falls is

- ✔ 41 degrees, 57 minutes, and 29.76 seconds north of the equator
- ✔ 121 degrees, 24 minutes, 34.73 seconds west of the prime meridian

Universal Transverse Mercator (UTM)

Universal Transverse Mercator is a modern coordinate system developed in the 1940s. It's similar to latitude and longitude, but it uses meters instead of degrees, minutes, and seconds. UTM coordinates are very accurate, and the system is pretty easy to use and understand.

Although the United States hasn't moved to the metric system, the system is widely used by GPS receivers. UTM coordinates are much easier than latitude and longitude to plot on maps. The two key values to convert metric measurements are

- ✔ **1 meter = 3.28 feet = 1.09 yards.**

    For ballpark measurements, a *meter* is a bit over a yard.

- ✔ **1 kilometer = 1,000 meters = 3,280 feet = 1,094 yards = 0.62 miles.**

    For ballpark measurements, a *kilometer* is a bit more than half a mile.

The UTM system is based on the simple A, B, C/1, 2, 3 coordinate system. The world is divided into *zones:*

- ✔ **60 primary zones run north and south.**

    Numbers identify the zones that run north and south.

- ✔ **20 zones run east to west.**

    These zones indicate whether a coordinate is in the Northern or Southern Hemisphere.

    Letters designate the east/west zones.

Sometimes the letter is dropped from a UTM coordinate, and only the zone is used to make things simpler. For example, because most of Florida is in Zone 17 R, if you were plotting locations in that state, you could just use Zone 17 in your UTM coordinates. Figure 2-6 shows the UTM zone map.

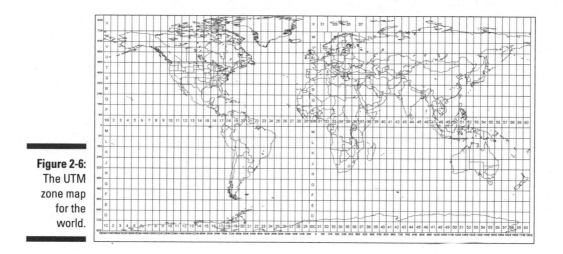

Figure 2-6:
The UTM
zone map
for the
world.

To provide a precise location, UTM uses two units:

✔ **Easting:** The distance in meters to the east from the start of a UTM zone line

   The letter *E* follows Easting values.

✔ **Northing:** The distance in meters from the equator

   The letter *N* follows Northing values.

   There's no such thing as a Southing. *Northing* is used in the Southern Hemisphere to describe the distance away from the equator, even though a location is south of the equator. (Is that weird, or what?)

Using my Dillon Falls example, if you use UTM to locate the falls, the coordinates look like this:

   10T 0627598E 4868251N

That means that the falls are in Zone 10T, which is 4,868,251 meters north of the equator and 627,598 meters east of where the zone line starts. (For those of you without a calculator in front of you, that's about 3,025 miles north of the equator, and about 390 miles east of where the Zone 10 line starts in the Pacific Ocean.)

Township and Range

The Township and Range coordinate system has been used since the 1790s to survey public lands in the United States. Technically, the official name of this system is the Public Land Rectangular Survey (PLS), but in practical use, most people call it Township and Range.

This coordinate system was developed after the American Revolution as a way to survey and grant title to land that was newly acquired following the country's independence. Thomas Jefferson helped develop the system, which was enacted under the Northwest Land Ordinance of 1785. Township and Range isn't used in the eastern United States (or in a few other states) because land surveys in those states had already been completed.

The system is based on the following components, which are shown in Figure 2-7:

- **Meridians and baselines.** These lines are the foundation of the Township and Range system:

    - *Meridians* are imaginary lines that run north to south.

    - *Baselines* are lines that run east to west.

    - An *initial point* is the spot where a meridian and a baseline meet.

        The California Bureau of Land Management has a nice online map of all the meridians and baselines at www.blm.gov/cadastral/ meridians/meridians.htm.

- **Townships:** *Townships* are the vertical part of the coordinate grid.

    - Each township is 6 miles in length, north to south.

    - Townships are identified by whole numbers starting with 1.

    - The first township at the intersection of a meridian and baseline is 1, the next township is 2, and so on.

    - If a township is north of the baseline, it's identified with an N; if it's south of the baseline, it's designated with an S. For example, the fifth township north of a meridian and baseline is T. 5 N.

- **Ranges:** *Ranges* are the hortizontal part of the grid scale.

    - Ranges are 6 miles in the east-west direction.

    - Ranges are numbered starting at the intersection of the meridian and the baseline.

    - In addition to a number, a range is identified as being east or west of a meridian. For example, the third range west of the meridian and baseline is R. 3 W.

The intersection of a township and range (a 36-square mile parcel of land) also is also called a township. This bit of semantics shouldn't have an effect on you using the coordinate system, but watch out for someone else getting his terminology confused.

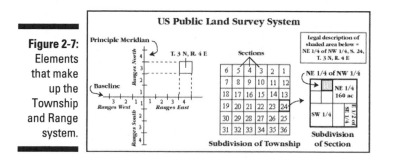

Figure 2-7: Elements that make up the Township and Range system.

Like other coordinate systems, Township and Range uses smaller measurement units to identify a precise location. These units include

✔ **Sections:** A 36-square mile township is further divided into 36 one-mile squares called *sections*.

Sections are numbered 1–36. Section 1 is in the top right of the township, and the numbers sequentially snake back and forth across the township, ending at 36 in the bottom right.

Specialized coordinate systems

Here are a few other coordinate systems, just so you know what they are:

MGRS (Military Grid Reference System): A coordinate system used by the U.S. and NATO military forces. It's an extension of the UTM system. It further divides the UTM zones into 100-kilometer squares labeled with the letters A–Z.

Proprietary grids: Anyone can invent a coordinate system for finding locations on a map. Examples of proprietary systems are ZIP code, the *Maidenhead Locator System* (a grid system for amateur radio operators), and *The Thomas Guides* (map books that match a location with a page number and grid).

State Plane Coordinate System: A coordinate system used in the United States. Each state contains at least one State Plane zone. Similar to the UTM system, it uses feet instead of meters.

USNG (United States National Grid): A coordinate system adopted by the Department of Homeland Security in 2005. Based on MGRS and developed for public safety use.

Most coordinate systems try to make navigation and surveying more accurate and simpler.

GPS is sending less-used coordinate systems the way of the dinosaur because you can quickly and easily get precise location positions in either UTM coordinates or latitude and longitude with an inexpensive GPS receiver.

> ✔ **Quarters:** Sections are divided even further into *quarters*.
>
> • Quarters are identified by the part of the section they occupy, such as northwest, northeast, southwest, or southeast.
>
> • You can further narrow the location with *quarter quarters* or *quarter quarter-quarters*.

Township and Range coordinates are a hodgepodge of abbreviations and numbers that lack the mathematical precision of latitude and longitude or UTM. For example, the Township and Range coordinates of Dillon Falls are

SE 1/4 of SW 1/4 of NE 1/4, Sect. 4, T. 19 S, R. 11 E, Willamette Meridian

To describe a location with this coordinate system, you start from the smallest chunk of land and then work your way up to larger chunks. Some people ignore this convention and reverse the order, skip the meridian, or use both halves and quarters. (Hey, it keeps life interesting. . . .)

Although scanned paper maps (such as USGS topographic maps) often show Township and Range information, most digital mapping software and GPS receivers don't support Township and Range. This is good news because latitude and longitude and UTM are much easier to use.

Township and Range information usually is omitted from digital maps because

✔ The coordinate system is difficult to mathematically model.

✔ Townships and sections may be oddly shaped because of previously granted lands, surveying errors, and adjustments for the curvature of the Earth.

Peter Dana's comprehensive Geographer's Craft Web site has lots of good technical information on coordinate systems:

```
www.colorado.edu/geography/gcraft/notes/coordsys/
            coordsys.html
```

Measuring Map Scales

Most maps have a *scale* — the ratio of the horizontal distance on the map to the corresponding horizontal distance on the ground. For example, one inch on a map can represent 1 mile on the ground.

The map scale is usually shown at the bottom of the map in the legend. Often, rulers with the scale mark specific distances for you. A scale from a USGS topographic map is shown in Figure 2-8.

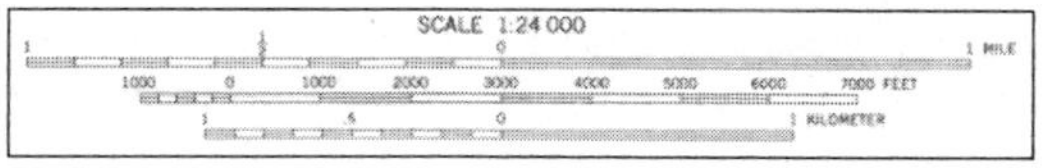

Figure 2-8:
Scale
information
in the
legend of a
USGS
topo-
graphic
map.

SCALE 1:24 000

1 MILE

1000   0   1000   2000   3000   4000   5000   6000   7000 FEET

1   5   0   1 KILOMETER

Many maps use a *representative fraction* to describe scale. This is the ratio of the map distance to the ground distance in the same units of measure. For example, a map that's 1:24,000 scale means that 1 inch measured on the map is equivalent to 24,000 inches (2,000 feet) on the ground. The number can be inches, feet, millimeters, centimeters, or some other unit of measure.

The units on the top and bottom of the representative fraction must be the same. You can't mix measurement units.

When you're dealing with scale, keep these guidelines in mind:

✔ **The smaller the number to the right of the 1, the more detail the map has.**

A 1:24,000 scale map has much more detail than a 1:100,000 scale map. A 1:24,000 map is a *large-scale* map, showing a small area.

✔ **The smaller the number to the right of the 1, the smaller the area the map displays.**

In Figure 2-9, the 1:100,000 scale map shows a much larger area than the 1:24,000 scale map. A 1:100,000 map is a *small-scale* map, showing a large area.

All sorts of rulers and measurement tools are calibrated to scale for measuring distance on paper maps. Mapping software makes distances easier and quicker to determine by offering tools that draw a line between two points and show an exact distance.

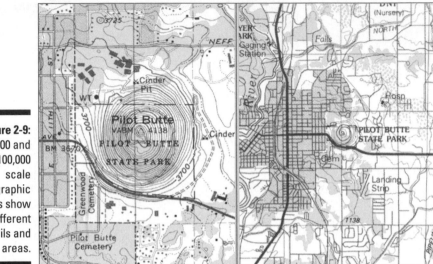

Figure 2-9:
1:24,000 and
1:100,000
scale
topographic
maps show
different
details and
areas.

Deciphering Map Symbols

Whether you're using a paper or a digital map, always familiarize yourself with the map's symbols. The more symbols you know, the better decisions you make when you're relying on a map for navigation. Some common symbols fount on USGS topographic maps are shown in Figure 2-10.

BUILDINGS AND RELATED FEATURES	
Building	■ □ ▨ ▨
School; church	⌐ ⌐
Built-up Area	
Racetrack	⬭ ⬭
Airport	✕ ✈
Landing strip	⊏⊐⊏⊐⊐
Well (other than water); windmill	○ ⌄
Tanks	● ◍
Covered reservoir	◍ ▨
Gaging station	⬟
Landmark object (feature as labeled)	○
Campground; picnic area	⌐ ⌐
Cemetery: small; large	[⌐] [Cem]

Figure 2-10:
Selected
USGS topo-
graphic map
symbols.

Mapmaker, mapmaker, make me a map

Data for digital maps comes from government sources and commercial mapmaking companies, which license their maps to other companies. These entities get data from satellites, aerial photographs, existing maps, or sometimes from folks driving and walking around with a GPS receiver, using a laptop to record data. As an example, here are the main providers of digital map data for the United States. (Most countries have a similar collection of government agencies and commercial sources that produce map data.)

✔ **Census Bureau:** The Census Bureau is in the map-making business because the data it collects is directly tied to location. Its Web site has information on its TIGER data.

   http://tiger.census.gov

✔ **FAA:** The Federal Aviation Administration National Aeronautical Charting Office (NACO) provides aviation charts and data for the United States.

   www.naco.faa.gov

✔ **Navteq:** Navigation Technologies is one of the largest commercial suppliers of street and road data. BMW, Garmin, Google, UPS, and many other companies and Web map sites use Navteq data. Nokia purchased the company in October 2007.

   www.navteq.com

✔ **NGA:** The National Geospatial-Intelligence Agency (formerly the National Imagery and Mapping Agency) is the Department of Defense agency for producing maps and charts of areas outside the United States.

   www.nga.mil

✔ **NOAA:** The National Oceanic and Atmospheric Administration creates and maintains nautical charts and other essential data for marine use.

   www.nauticalcharts.noaa.gov

✔ **Tele Atlas:** The other big player in digital street data for GPS companies, car navigation systems, and Web map sites. Acquired by GPS manufacturer TomTom in November 2007.

   www.teleatlas.com

✔ **USGS:** The United States Geological Survey (USGS) has been making U.S. land maps since 1879.

   http://topomaps.usgs.gov

✔ **States, counties, and cities:** States, counties, and larger cities have Geographic Information System (GIS) divisions that produce maps. This data is often available to the public free or for a small cost. To find whether digital map data is available for your area, search Google for your state, county, or city and *GIS.*

Map symbols aren't universal. A symbol can have different meanings on different maps. For example, the symbol for a secondary highway on a USGS topographic map is a railroad on a Swiss map.

These Web sites show what symbols for different types of maps mean:

- ✔ Topographic maps

  `http://erg.usgs.gov/isb/pubs/booklets/symbols`
- ✔ Aeronautical charts

  `www.naco.faa.gov/index.asp?xml=naco/online/aero_guide`
- ✔ Nautical charts

  `http://chartmaker.ncd.noaa.gov/mcd/chart1/chart1hr.htm`

Digital Map Data

Here I concentrate on digital map data that the U.S. government makes available free. You can make your own maps with this data by using freeware and shareware mapping programs, which I discuss in Part III.

Although other countries use the same or similar data formats as the U.S. government for producing digital maps, trying to obtain detailed international map data can be difficult. In some parts of the world, the government tightly controls maps and map data because they're considered a key part of a country's national security. Chapter 20 lists a number of Internet resources where you can get maps and map data for areas outside the United States.

TIGER

The U.S. Census Bureau produces Topologically Integrated Geographic Encoding and Referencing (TIGER) data for compiling maps with demographic information. This vector data is a primary source for creating digital road maps of the United States. Figure 2-11 shows a map of downtown San Francisco created from Census Bureau TIGER data.

TIGER data is free (check out the Census Bureau Web site at `www.census.gov/geo/www/tiger/index.html`), but in some areas the data isn't very accurate; roads don't appear on the map and addresses aren't in the right locations. The government is always improving the accuracy of the dataset, but you're better off using some of the free and commercial street map programs and Web sites discussed in Chapters 13 and 18.

Raster and vector maps

Raster maps are composed of a series of pixels (picture elements) aligned in a grid. Each pixel (or bit) contains information about the color to be displayed or printed. Because bitmap images can be very large, graphics formats such as GIF (Graphic Interchange Format) and TIFF (Tag Image File Format) use compression to make images smaller. An example of a raster map is a paper map that's been scanned and saved as graphic file. Raster maps tend to show more detail than vector maps and don't look as computer-generated. Figure 2-1 shows a raster map.

Vector maps are composed of many individual objects. The objects are mathematically based. They can be points, lines, and shapes. Each object has properties that define its appearance and attributes, such as color, thickness, and style. If you've used such graphics software as Adobe Illustrator or CorelDRAW (called *object-oriented* drawing programs), you've used a vector graphics program. A map can contain thousands of these objects, but vector maps tend to be smaller than raster maps because it's more efficient to describe an object mathematically than draw it as a bitmap. Vector maps are *scalable,* which means you can resize a vector map without distorting the map's information. Vector maps appear as if they were created with lines and shapes. Figure 2-2 shows a vector map.

Figure 2-11: TIGER data viewed from the Census Bureau's Web-based map viewer.

Digital Line Graph (DLG)

Digital Line Graph (DLG) data from the USGS is used to create vector maps. The data includes transportation networks, *hydrography* (surface water measurements), boundaries, elevation contours, and man-made features. The format is similar to TIGER data but generally has more roads and features that are more accurate.

Figure 2-12 shows a map of Wassaw Sound, Georgia, created with USGS DLG data. (Wassaw Sound, just outside Savannah, was the yachting venue for the Atlanta Olympics in 1996.)

DLG data has basic information such as road types, bridges, and highway route numbers, but only GIS software can easily access these attributes.

Figure 2-12:
A map
created with
DLG data.

Elevation data

You're likeliest to run into one of these main data formats used to represent elevation:

- **Digital Elevation Model (DEM):** The raster data format the USGS uses to record elevation information (based on topographic maps) and create three-dimensional representations of the terrain. Figure 2-13 shows a map of Mount Bachelor, Oregon, generated from Digital Elevation Model data with the 3DEM mapping program.

- **National Elevation Dataset (NED):** This format shows digital elevation data in shaded relief. It's designed for seamless coverage of the United States in large raster files and is available in 10- and 30-meter resolution.

- **Shuttle Radar Topography Mission (SRTM):** Very accurate, 30-meter elevation data derived from Space Shuttle radar images.

Figure 2-13:
A map created with DEM data.

Digital Raster Graphics (DRG)

Digital Raster Graphics (DRG) data is a scanned image of a USGS topographic map. These digital maps are available free on the Internet or sold commercially in collections on CDs or DVDs. Figure 2-14 shows a DRG map from the TerraServer-USA Web server. This map is a digital version of a 1:24,000 scale topographic map.

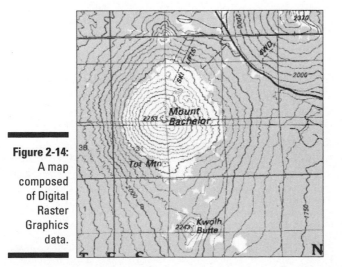

Figure 2-14:
A map composed of Digital Raster Graphics data.

Mr. Sid and déjà vu

In your travels to the world of maps you may encounter Mr. SID and experience déjà vu (no, you're not in Oz, Toto). MrSID is the Multi-Resolution Seamless Image Database — a file format used for distributing large images over networks, originally developed by a company called LizardTech. Graphics in MrSID format are compressed with a *lossless compression algorithm* (a method of compressing data that guarantees the original data can be restored exactly) designed to produce relatively small, high-resolution images. DjVu (pronounced

"déjà vu") is another digital document format with advanced compression technology.

These file formats are perfect for aerial and satellite images that have large file sizes, and the government is increasingly using them for distributing data (especially scanned historic maps and documents). A number of free viewers and browser plugins support MrSID and DjVu; use Google to find download sites. (One of my favorite standalone viewers is IrfanView, which is available at `www.irfanview.com`.)

These digital maps are scanned at 250 dpi (dots per inch) and stored in a TIFF file format, using embedded GeoTIFF (geographic information) tags for location data.

You can view the map by itself or view both the map and its location data. Use one of the following methods:

✔ View the map by opening the DRG file with any current graphics program that supports large TIFF files.

✔ Use the DRG file with a mapping program that supports GeoTIFF to view the map and access its location data.

For more technical details about USGS digital map data, check out the agency's product Web site at `http://topomaps.usgs.gov/drg`.

Digital Orthophoto Quadrangle (DOQ)

Digital Orthophoto Quadrangle (DOQ) data consists of a computer-generated image of an aerial photograph. The image is corrected so that camera tilt and terrain relief don't affect the accuracy. DOQs combine the image characteristics of a photograph with the geometric qualities of a map. Figure 2-15 shows a DOQ map of Mount Bachelor, Oregon, from the TerraServer-USA Web site.

The USGS has DOQs available for the entire United States. Most are grayscale, infrared photos; there are higher-resolution color photos for a few large U.S. metropolitan areas.

There is a booming business providing high-resolution, color aerial photo-graphs to individuals, government agencies, corporations, and educational and non-profit organizations. Companies like AirPhotoUSA (www.airphotousa.com), Aerial Express (http://aerialsexpress.com) sell imagery with quality and resolution that's close to what was only available to intelligence agencies. If you want aerial photographs for business or government pur-poses, check these commercial sources.

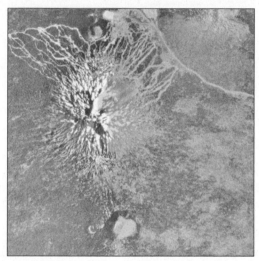

Figure 2-15:
A Digital
Orthophoto
Quadrangle
map.

Satellite data

Satellites are the most exotic source of data for digital maps. Orbiting several hundred miles above the Earth, satellites provide photographs and other sensor data. NASA's Landsat, the Space Shuttle, and other satellites collect raster data for most of the Earth. It's available both free and commercially.

Resolution defines the smallest object a satellite can distinguish. A satellite with 1-meter resolution can distinguish objects down to a meter (a little under 40 inches) in size.

Figure 2-16 is a 10-meter resolution SPOT satellite data of Mount Bachelor. (SPOT, which stands for Systeme Pour l'observation de la Terre, is a French commercial satellite program that started in 1986.) This used to be state of the art, but times have changed. Compare this to the images shown in Chapter 15.

I spy

Digital Globe (www.digitalglobe.com) is the leading commercial provider of satellite imagery. QuickBird, its first satellite launched in 2001, is capable of taking black-and-white photos with 0.61-meter resolution and color images at around 2.5 meters. The company's WorldView-1 satellite, which started acquiring images in October 2007, can snap 0.5-meter resolution color photos (construction of WorldView-2 is under way, with the goal of having three high-resolution, Earth observation satellites in orbit by 2008/2009).

Because the WorldView image quality is so good, Digital Globe is supplying data to the super secret National Geospatial-Intelligence Agency. (The best government spy satellites, by the way, are thought to have a resolution of 0.1 meters and reputedly discern objects about the size of a baseball. Despite what you see in the movies, license plate numbers are too small to be read by spy satellites, especially considering they don't face toward space.)

Just a few years ago, access to good satellite imagery was pretty limited — unless you worked for a government agency or shelled out some hard-earned cash of your own. That's changed now, with Google dramatically increasing the availability of free, high-resolution color imagery through its Web-based Google Maps and standalone Google Earth applications. I discuss both of these products in Chapters 15 and 18.

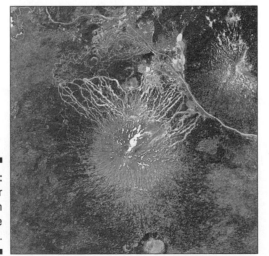

Figure 2-16:
A 10-meter resolution satellite image.

Part II
All About GPS

In this part . . .

GPS stands for Global Positioning System, which is a nifty satellite system that tells you your location anywhere on planet earth (with a few limitations). This part is all about GPS. We start with a broad overview of the satellite system and then work back down to earth and discuss handheld, consumer GPS receivers. By the end of the part, you'll understand the technology behind GPS (without needing to be a rocket scientist), be able to evaluate the jungle of GPS receivers on the market, and have the skills and knowledge to make practical use of a GPS receiver without being intimidated or confused by it.

Chapter 3

GPS Fundamentals

ou've heard about GPS and probably know that it has something to do with electronic gadgets and satellites that tell you where you're located. That's great for starters, but you need a more detailed understanding to use GPS. This chapter takes you through some of the fundamentals of GPS so you have a better grasp of what it is and how you use it.

Global Position What?

GPS stands for *Global Positioning System*. A special radio receiver measures the distance from your location to satellites that orbit the earth broadcasting radio signals. GPS can pinpoint your position anywhere in the world. Pretty cool, huh? Aside from buying the receiver, the system is free for anyone to use.

You can purchase an inexpensive GPS receiver, pop some batteries in it (or plug it into your car's cigarette lighter), turn it on, and presto! Your location appears on the screen. No paper map, compass, sextant, or sundial is required. Just like magic. It's not really magic though; it's evolved from some great practical applications of science that have come together over the last 50 or so years.

Other satellite Global Positioning Systems are either in orbit or planned by other countries, but this book uses the term *GPS* for the Global Positioning System operated by the U.S. government.

A short history of GPS

Military, government, and civilian users all over the world rely on GPS for determining their location, but using radio signals for navigation isn't new, and has been around since the 1920s. Before GPS, LORAN (Long Range Aid to Navigation), a position-finding system that measured the time difference of arriving radio signals, was developed during World War II.

The foundations of GPS date back to 1957 when the Russians launched *Sputnik,* the first satellite to orbit the Earth. Sputnik used a radio transmitter to broadcast telemetry information. Scientists at the Johns Hopkins Applied Physics Lab discovered that the Doppler shift phenomenon applied to the spacecraft — and almost unwittingly struck gold.

A down-to-earth, painless example of the Doppler shift principle is when you stand on a sidewalk and a police car speeds by in hot pursuit of a stolen motorcycle. The pitch of the police siren increases while the car approaches you and then drops sharply after it passes by.

American scientists figured out that if they knew the satellite's precise orbital position, they could accurately locate their exact position on Earth by listening to the pinging sounds and measuring the satellite's radio signal Doppler shift. Satellites offered some possibilities for a navigation and positioning system, and the U.S. Department of Defense (DoD) explored the concept.

By the 1960s, several rudimentary satellite-positioning systems existed. The U.S. Army, Navy, and Air Force were all working on independent versions of radio navigation systems that could provide accurate positioning and all-weather, 24-hour coverage. In 1973, the Air Force was selected as the lead organization to consolidate all the military satellite navigation efforts into a single program. This evolved into the NAVSTAR (contrary to public belief, NAVSTAR is just a name and not an acronym) Global Positioning System, which is the official name for the United States' GPS program.

The U.S. military wasn't just interested in GPS for navigation. A satellite location system can be used for weapons-system targeting. Smart weapons, such as the Tomahawk cruise missile, use GPS in their precision guidance systems. GPS, combined with contour-matching radar and digital image-matching optics, makes a Tomahawk an extremely accurate weapon. The possibility of an enemy using GPS against the United States is one reason why civilian GPS receivers are less accurate than their restricted-use military counterparts.

The first NAVSTAR satellite was launched in 1974 to test the concept. By the mid-1980s, more satellites were put in orbit to make the system functional, and it was opened to civilian use. In 1994, the planned full constellation of 24 satellites was in place. Soon, the military declared the system completely operational. The program is wildly successful and is still funded through the U.S. DoD.

GPS Deconstructed

The intricacies of GPS are steeped in mathematics, physics, and engineering, but you don't need to be a rocket scientist to understand how GPS works. GPS is composed of three components (as shown in Figure 3-1):

✔ Satellites (space segment)

✔ Ground stations (control segment)

✔ Receivers (user segment)

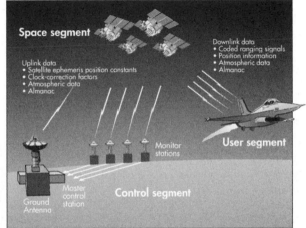

Figure 3-1:
The building
blocks of
GPS.

Eyeing satellites

In GPS jargon, satellites are the space segment of GPS; that is, the components that are out in space. A minimum of 24 GPS satellites (called a constellation) orbit about 12,000 miles above the Earth (as shown in Figure 3-2). The satellites zoom through the heavens at around 7,000 miles per hour. It takes about 12 hours for a satellite to orbit the Earth, passing over the exact same spot approximately every 24 hours. The satellites are positioned so a GPS receiver can receive signals from at least six of the satellites at any time, at any location on Earth (if nothing obstructs the signals).

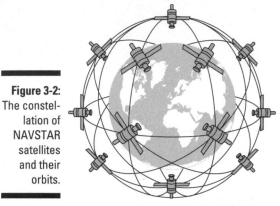

Figure 3-2:
The constel-
lation of
NAVSTAR
satellites
and their
orbits.

A satellite has three key pieces of hardware:

- ✔ **Computer:** This onboard computer controls a satellite's flight and other functions.
- ✔ **Atomic clock:** This keeps accurate time within 3 *nanoseconds* (around three-billionths of a second).
- ✔ **Radio transmitter:** This sends signals to Earth.

GPS satellites don't just help you stay found. All GPS satellites since 1980 carry NUDET sensors. No, this isn't some high-tech pornography-detection system. NUDET is an acronym for NUclear DETonation; GPS satellites have sensors to detect nuclear-weapon explosions, assess the threat of nuclear attack, and help evaluate nuclear strike damage.

The solar-powered GPS satellites have a limited life span (around 10 years). When they start to fail, spares are activated or new satellites are sent into orbit to replace the old ones. This gives the government a chance to upgrade the GPS system by putting hardware with new features into space.

During the spring of 2008, there were 32 broadcasting GPS satellites in orbit, making the system more precise, reliable, and available.

The GPS network is vulnerable to extreme sun activity. For two days in December 2006, the GPS system was disrupted by a solar flare that produced strong radio waves preventing a large number of GPS receivers from receiving satellite signals. The next predicted period of strong solar activity is in 2011. Be sure to have a compass and paper map ready as a backup, just in case.

GPS radio signals

GPS satellites transmit two types of radio signals: C/A-code and P-code. Briefly, here are the uses and differences of these two signal types.

Coarse Acquisition (C/A-code)

Coarse Acquisition (C/A-code) is the type of signal that consumer GPS units receive. C/A-code is sent on the L1 band at a frequency of 1575.42 MHz.

C/A broadcasts are known as the Standard Positioning Service (SPS).

C/A-code is less accurate than P-code (see the following section) and is easier for U.S. military forces to jam and *spoof* (broadcast false signals to make a receiver think it's somewhere it's really not).

The advantage of C/A-code is that it's quicker to use for acquiring satellites and getting an initial position fix. Some military P-code receivers first track on the C/A-code and then switch to P-code.

Precision (P-code)

P-code provides highly precise location information. P-code is difficult to jam and spoof. The U.S. military is the primary user of P-code transmissions, and it uses an encrypted form of the data (Y-code) so that only special receivers can access the information. The P-code signal is broadcast on the L2 band at 1227.6 MHz.

P-code broadcasts are known as the Precise Positioning Service (PPS).

Covering ground stations

Ground stations are the control segment of GPS (refer to Figure 3-1). Five unmanned ground stations around the Earth monitor the satellites. Information from the stations is sent to a master control station — the Consolidated Space Operations Center (CSOC) at Schriever Air Force Base in Colorado — where the data is processed to determine each satellite's ephemeris and timing errors.

An *ephemeris* is a list of the predicted positions of astronomical bodies, such as the planets or the Moon. Ephemerides (the plural of *ephemeris*) have been around for thousands of years because of their importance in celestial navigation. Ephemerides are compiled to track the positions of the numerous satellites orbiting the earth.

The processed data is sent to the satellites once daily with ground antennas located around the world. This is kind of like synchronizing a personal digital assistant (PDA) with your personal computer to ensure that all the data between the two devices matches. Because the satellites have small built-in rockets, the CSOC can control them to ensure that they stay in a correct orbit.

GPS receivers

Anyone who has a GPS receiver (the user segment; refer to Figure 3-1) can receive the satellite signals to determine where he or she is located.

Satellite data

GPS units receive two types of data from the NAVSTAR satellites.

Almanac

Almanac data contains the approximate positions of the satellites. The data is constantly being transmitted and is stored in the GPS receiver's memory.

Ephemeris

Ephemeris data has the precise positions of the satellites. To get an accurate location fix, the receiver has to know how far away a satellite is. The GPS receiver calculates the distance to the satellite by using signals from the satellite.

Using the formula *Distance = Velocity x Time*, a GPS receiver calculates the satellite's distance. A radio signal travels at the speed of light (186,000 miles per second). The GPS receiver needs to know how long the radio signal takes to travel from the satellite to the receiver in order to figure the distance. Both the satellite and the GPS receiver generate an identical pseudo-random code sequence. When the GPS receiver receives this transmitted code, it determines how much the code needs to be shifted (using the Doppler-shift principle) for the two code sequences to match. The shift is multiplied by the speed of light to determine the distance from the satellite to the receiver.

Multiple satellites

A GPS receiver needs several pieces of data to produce position information:

- **Location:** A minimum of three satellite signals are required to find your location.

- **Position:** Four satellite signals are required to determine your position in three dimensions: latitude, longitude, and elevation.

Receiver types

All sorts of GPS receivers are on the market, but they tend to fall into five general categories.

Consumer models

Consumers can buy practical GPS receivers in electronics and sporting goods stores and from Web sites. They're easy to use and are mostly targeted for recreational and other uses that don't require a high level of location precision. Handheld consumer GPS receivers are reasonably priced, from less than $100 to $400 in the United States. Auto navigation systems range from $150 to over $1,000 (with the higher-end model prices falling).

This book discusses the features and use of the consumer-oriented, handheld, and automotive types.

Channels refer to the number of GPS satellites a receiver can simultaneously monitor. Older models have 8 channels, whereas contemporary receivers have 12 or more. Models with more than 12 channels typically acquire satellites quicker and work better under foliage and in city "urban canyons."

U.S. military/government models

GPS units that receive P-code and Y-code are available only to the government.

Mapping/resource models

These portable receivers collect location points and line and area data that can be input into a Geographic Information System (GIS). They are more precise than consumer models, can store more data, and are much more expensive.

Survey models

These are used mostly for surveying land, where you may need accuracy down to the centimeter for legal or practical purposes. These units are extremely precise and store a large amount of data. They tend to be large, complex to use, and very expensive.

Commercial transportation models

These GPS receivers, not designed to be handheld, are installed in aircraft, ships, boats, trucks, and cars. They provide navigation information appropriate to the mode of transportation and can send the vehicle's location to a monitoring facility.

Trimble Navigation (www.trimble.com) is one of the biggest players in the professional GPS receiver market. If you're interested in discovering the workings of commercial and higher-end GPS units and their features, check out the Trimble Web site.

GPS receiver accuracy

According to the government and GPS receiver manufacturers, expect your GPS unit to be accurate within 49 feet (that's 15 meters for metric-savvy folks). If your GPS reports that you're at a certain location, you can be reasonably sure that you're 49 feet or less from that exact set of coordinates.

GPS receivers tell you how accurate your position is. Based on the quality of the satellite signals that the unit receives, the screen displays the estimated accuracy in feet or meters. Accuracy depends on

✔ Receiver location

✔ Number of satellite signals being received

✔ Obstructions that block satellite signals

Even if you're not a U.S. government or military GPS user, you can get more accuracy by using a GPS receiver that supports enhanced location data. This information is broadcast over radio signals that come from either

✔ Non-GPS satellites

✔ Ground-based beacons

Selective Availability (SA)

The average GPS user didn't always have 15-meter accuracy. In the 1970s, studies showed that the less-accurate C/A-code for nonmilitary use was more accurate than what the U.S. government intended. Originally thought to provide accuracy within 100 meters, experiments showed that C/A accuracy was in the range of 20–30 meters. To reduce the accuracy of C/A-code, the U.S. government developed *Selective Availability (SA).* SA adds errors to NAVSTAR satellite data and prevents consumer GPS receivers from providing an extremely precise location fix.

Selective Availability was temporarily turned off in 1990 during the Persian Gulf War. There weren't enough U.S. and allied military P-code GPS receivers, so the Coalition troops used civilian GPS receivers. The Gulf War was the first use of GPS in large-scale combat operations.

On May 2, 2000, SA was turned off permanently. Overnight, the accuracy of civilian GPS users went from 100 meters to 15 meters. Turning off SA on a global scale was directly related to the U.S. military's ability to degrade the C/A-code on a regional basis. For example, during the invasion of Afghanistan, the American military jammed GPS signals in Afghanistan to prevent the Taliban from using consumer receivers in operations against American forces.

Two common sources of more accurate location data are

- Differential GPS (DGPS)
- Wide Area Augmentation System (WAAS)

I cover both DGPS and WAAS in the following section on GPS receiver features.

Table 3-1 shows the accuracy you can expect from a GPS receiver. These numbers are guidelines; at times, you may get slightly more or less accuracy.

Table 3-1	GPS Accuracy	
GPS Mode	*Distance in Feet*	*Distance in Meters*
GPS without SA	49	15
GPS with DGPS	10–16	3–5
GPS with WAAS	10	3

Although survey-grade GPS receivers can provide accuracy of less than 2 centimeters, they are very specialized and expensive, require a lot of training, and aren't very portable. Their accuracy is achieved with DGPS and post-processing collected data to reduce location errors. The average GPS user doesn't need this level of precision.

Clouds, rain, snow, and weather don't reduce the strength of GPS signals enough to reduce accuracy. The only way that weather can weaken signals is when a significant amount of rain or snow accumulates on the GPS receiver antenna or on an overhead tree canopy.

Stuff Your GPS Receiver Can Tell You

GPS receivers provide your location and other useful information:

- **Time:** A GPS receiver receives time information from atomic clocks, so it's much more accurate than the average wristwatch.
- **Location:** GPS provides your location in three dimensions:
  - Latitude (x coordinate)
  - Longitude (y coordinate)
  - Elevation

The vertical (elevation) accuracy of consumer GPS receivers isn't that great. It can be within 15 meters, 95 percent of the time. Some GPS units incorporate more accurate barometric altimeters for better elevation information.

Your location can be displayed in a number of coordinate systems, such as

- Latitude/longitude

- Universal Transverse Mercator (UTM)

I cover coordinate systems in Chapter 2.

- **Speed:** When you're moving, a GPS receiver displays your speed. (And if you've entered a waypoint, an Estimated Time of Arrival based on your current speed.)

- **Direction of travel:** A GPS receiver can display your direction of travel if you're moving. It can also provide guidance on how to reach a set location.

    If you're stationary, the unit can't use satellite signals to determine which direction you're facing.

    Whether you're moving or standing still, some GPS units have electronic compasses that show the direction the receiver is pointed.

- **Stored locations:** With a GPS receiver, you can store locations where you've been or want to go. These location positions are *waypoints*. Waypoints are important because a GPS unit can supply you with directions and information on how to get to a waypoint. A collection of waypoints that plots a course of travel is a *route,* which can also be stored. GPS receivers also store *tracks* (which are like an electronic collection of breadcrumb trails that show where you've been).

    In Chapter 4, I show how to create and use waypoints, routes, and tracks.

- **Maps:** Most GPS receivers can display some form of a map (what appears on the map and how much detail depends on the model).

- **Cumulative data:** A GPS receiver can also keep track of travel information, such as the total distance traveled, average speed, maximum speed, minimum speed, elapsed time, and time to arrival at a specified location.

All this information displays on different pages of the GPS receiver's display screen. One page shows satellite status, another page displays a map, another displays trip data, and so on. With buttons or other controls on the receiver, you can scroll to an information page to view the data that you're interested in seeing.

GPS errors

A number of conditions can reduce the accuracy of a GPS receiver. From a *top-down* perspective (from orbit to ground level), the possible sources of trouble look like this:

✔ **Ephemeris errors:** Ephemeris errors occur when the satellite doesn't correctly transmit its exact position in orbit.

✔ **Ionosphere conditions:** The *ionosphere* starts at about 43–50 miles above the Earth and continues for hundreds of miles. Satellite signals traveling through the ionosphere are slowed down because of *plasma* (a low-density gas). Although GPS receivers attempt to account for this delay, unexpected plasma activity can cause calculation errors.

✔ **Troposphere conditions:** The *troposphere* is the lowest region in the Earth's atmosphere and goes from ground level up to about 11 miles. Variations in temperature, pressure, and humidity all can cause variations in how fast radio waves travel, resulting in relatively small accuracy errors.

✔ **Timing errors:** Because placing an atomic clock in every GPS receiver is impractical, timing errors from the receiver's less-precise clock can cause slight position inaccuracies.

✔ **Multipath errors:** When a satellite signal bounces off a hard surface (such as a building or canyon wall) before it reaches the receiver, a delay in the travel time occurs, which causes an inaccurate distance calculation.

✔ **Poor satellite coverage:** When a significant part of the sky is blocked, your GPS unit has difficulty receiving satellite data. Poor placement of a receiver in a car, building interiors, streets surrounded by tall buildings, parking garages, dense tree canopies, canyons, and mountainous areas are typical problem areas. A satellite that provides a good signal one day may provide a poor signal at the exact same location on another day because its position has changed and a tree is now blocking its signal. The more open sky you have, the better your chances of not having satellite signals blocked.

If satellite coverage is poor, try moving to a different location to see whether you get any improvement.

GPS Receiver Features

A number of GPS receivers are on the market, and they sport various features. Here are the more common ones (keep in mind that usually a GPS receiver with more features costs more).

GPS manufacturers have done a pretty good job making user interfaces fairly easy to use. After you know the basic concepts of GPS receivers and are familiar with a manufacturer's user interface, a GPS unit is usually as easy to use as a cellphone and certainly easier to use than a personal computer.

Chipset

In the old days, most GPS receivers were created equal when it came to sensitivity and the time it took to get a satellite fix. That's changed with advances in chip technology that provide faster initial position fixes and increased sensitivity under tree canopies and in canyons (either those in the wilderness or those formed by tall buildings in a city). SiRF Technology Holdings and Broadcom (formerly Global Locate) both supply GPS manufacturers with high-sensitivity chips. Older GPS units that don't have these latest chips still work fine. They're just a little slower and less sensitive than their younger relatives.

In addition to the GPS chip, a receiver also has a processing chip for displaying graphics and performing calculations. Just like a computer, models with more powerful chips perform a bit faster. (Unfortunately, GPS manufacturers don't advertise chip speeds like PCs, so you need to compare different models to determine performance.) Proprietary algorithms inside the chips used for computing location play a significant role in GPS receiver performance.

Display and output

GPS receivers display information or output data in one of three ways:

- **Monochrome screen:** It used to be that most GPS receivers had monochrome liquid crystal display (LCD) screens. That's changing, and monochrome GPS units are going the way of black-and-white televisions.

- **Color screen:** Living color is where it's at these days. Color is much more readable and manufacturers are boasting bigger screens, more pixels, and higher resolution.

- **No screen:** Some GPS receivers only transmit data through an expansion slot, cable, or wireless Bluetooth connection; a receiver with a cable is often called a *mouse* GPS receiver because it resembles a computer mouse. Such receivers are designed to interface with a laptop computer or PDA running special software. Figure 3-3 shows a DeLorme Earthmate GPS unit attached to a laptop. All GPS data is sent to the laptop and processed there with mapping software. A small handheld GPS receiver is shown on top of the laptop for size comparison.

    Most handheld GPS receivers that have screens can also output data to a PC or PDA with a serial or USB cable.

A GPS receiver's screen size depends on the receiver's size. Smaller, lighter models have small screens; larger units sport bigger screens.

Figure 3-3:
A GPS
receiver
without a
screen.

Generally, a bigger screen is easier to read. Different models of GPS receiver also have different pixel resolutions — the higher the screen resolution, the crisper the display will be. However, on some of the newer handheld models, high-resolution screens can actually be more difficult to read in direct sunlight than low-resolution screens. For night use, all screens can be backlit.

Alarms

Some GPS receivers can transmit a tone or display a message when you approach a location that you specify. This proximity alarm feature can be especially useful if you're hiking and trying to find a place when visibility is limited by darkness or inclement weather — or you're busy doing something else and aren't looking at the GPS receiver screen.

Built-in maps

Every handheld GPS receiver has an information page that shows waypoints and tracks (for more on waypoints and tracks, see Chapter 4). The page is a simple map that plots travel and locations. It doesn't show roads, geographic features, or man-made structures. Figure 3-4 shows three handheld GPS receiver screens: a simple location plotting on the left, a basemap on the right, followed by a detailed, uploaded map.

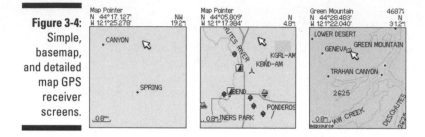

Figure 3-4:
Simple,
basemap,
and detailed
map GPS
receiver
screens.

Many handheld receivers can display maps that show roads, rivers, cities, and other features on their screens. (All automotive GPS units have maps. I talk about these receivers in Chapter 6.) You can zoom in and out to show different levels of detail. Two types of maps are associated with mapping GPS units:

✔ **Basemap:** Most GPS units have a basemap loaded into read-only memory that contains roads, highways, water bodies, cities, airports, railroads, and interstate exits.

✔ **Uploadable map:** More detailed maps can be purchased and added to mapping units (either by uploading to internal memory or by inserting a memory card). You can install road maps, topographic maps, and nautical charts. Many of these maps also have built-in databases, so your GPS receiver can display restaurants, gas stations, or attractions near a certain location. Models that can display certain types of aerial and satellite photos are also becoming available.

Chapter 11 covers how to use GPS receiver manufacturer mapping programs to upload maps to your GPS unit.

Refer to Figure 3-4 to see screens from a GPS receiver with a simple plot map, a basemap, and a detailed, uploaded map.

GPS receivers that display maps use proprietary map data from the manufacturer. That means you can't load another manufacturer's or software company's maps into a GPS receiver. However, clever hackers reverse-engineered Garmin's map format. You can download programs from the Internet to create and upload your maps to Garmin GPS receivers; GPSmapper is one such popular tool.

A handheld GPS receiver's screen is only several inches across. The limitations of such a small display certainly don't make the devices replacements for traditional paper maps.

Enhanced accuracy

Many GPS receivers have features that allow you to increase the accuracy of your location by using radio signals not associated with the GPS satellites (this is *augmentation* in GPS-geek talk). If you see that a GPS receiver supports WAAS or Differential GPS, it has the potential to provide you with more accurate location data.

WAAS

Wide Area Augmentation System (WAAS) combines satellites and ground stations for position accuracy of better than 3 meters. Vertical accuracy is also improved to 3 to 7 meters.

WAAS is a Federal Aviation Administration (FAA) system, designed so GPS can be used for airplane flight approaches. The system has a series of ground-reference stations throughout the United States (new stations are planned for Canada and Mexico that will increase signal availability). These stations monitor GPS satellite data and then send the data to two master stations — one on the west coast and the other on the east coast. These master stations create a GPS message that corrects for position inaccuracies caused by satellite orbital drift and atmospheric conditions. The corrected messages are sent to non-NAVSTAR satellites in stationary orbit above the equator. The satellites then broadcast the data to GPS receivers that are WAAS-enabled.

GPS units that support WAAS have a built-in receiver to process the WAAS signals. You don't need more hardware. Some handheld GPS receivers support turning WAAS on and off. If WAAS is on, battery life is shorter (although not as significantly as when you're using the backlight). In fact, on these models, you can't use WAAS if the receiver's battery-saver mode is activated. Whether you turn WAAS on or off depends on your needs. Unless you need a higher level of accuracy, you can leave WAAS turned off if your GPS receiver supports toggling it on and off. WAAS is ideally suited for aviation as well as for open land and marine use. However, the system may not provide any benefit in areas where trees or mountains obstruct the view of the horizon.

Under certain conditions — say, when weak WAAS satellite signals are being received or the GPS receiver is a long way from a ground station — accuracy can actually worsen when WAAS is enabled.

WAAS is only available in North America. Other governments are establishing similar systems that use the same format radio signals such as

- ✔ European Euro Geostationary Navigation Overlay Service (EGNOS)
- ✔ Japanese Multi-Functional Satellite Augmentation System (MTSAS)

Differential GPS

Surveyors and other workers that demand a high level of precision use Differential GPS (DGPS) to increase the position accuracy of a GPS receiver. A stationary receiver measures GPS timing errors and broadcasts correction information to other GPS units that are capable of receiving the DGPS signals. Consumer GPS receivers that support DGPS require a separate beacon receiver that connects to the GPS unit. Consumers can receive DGPS signals from free or commercial sources.

Unless you're doing survey or other specialized work, you really don't need DGPS capabilities. For consumer use, the increased accuracy of DGPS has mostly been replaced with WAAS.

Coast Guard DGPS

DGPS signals are freely broadcast by a series of U.S. Coast Guard stations in the United States. Whether you can receive these Coast Guard broadcasts depends on your location.

For more information on DGPS, including coverage maps, pay a visit to www. navcen.uscg.gov/dgps/coverage/default.htm.

Commercial DGPS

DGPS services are offered commercially for the surveying market. You can rent or purchase electronic and radio equipment for gathering precise location information in a relatively small area.

Antennas

Well, yes, a GPS unit has to have an antenna to receive radio signals to do you any good. Several types are available, each with its advantages.

Internal antennas

All GPS receivers have one of two kinds of built-in antennas. One antenna design isn't superior to the other; performance is related to the receiver's antenna size. (Cough . . . bigger is better.) Automotive GPS receivers may have antennas that flip up or that are built into the unit.

Patch

An internal *patch antenna* is a square conductor mounted over a *groundplane* (another square piece of metal). Patch antenna models reacquire satellites faster after losing the signal.

For best performance with an internal patch antenna, hold the handheld receiver face up and parallel with the ground.

Quad helix

An internal *quadrifilar helix antenna* (or *quad helix)* is a circular tube wrapped with wire. Quad helix antennas are more sensitive and work better under tree cover than the other types. (Antennas that protrude from a GPS receiver's case typically indicate a quad helix antenna.)

For best performance with an internal quad helix antenna, hold the handheld GPS receiver so that the top is pointing to the sky.

External antennas

Some GPS receivers have connectors for attaching external antennas. An external antenna is useful if the GPS receiver's view of the sky is otherwise blocked, like in a boat, a car, an airplane, or buried deep in a backpack.

Reradiating antennas

If a GPS receiver doesn't have a jack for connecting an external antenna, you can improve the reception with a *reradiating antenna.* These antennas work just as well as conventional external antennas that plug into a GPS receiver.

A reradiating antenna combines two GPS antennas:

- ✓ One antenna receives the GPS signal from the satellites.
- ✓ The other antenna is connected to the first and positioned next to the GPS unit's internal antenna.

Here are a couple of sources for reradiating antennas:

- ✓ **Roll your own:** If you're handy with a soldering iron, search Google for *reradiating antenna GPS* for tips on how to make one yourself.
- ✓ **Buy one:** Purchase an assembled reradiating antenna from Pc-Mobile at www.pc-mobile.net/gpsant.htm.

Internal storage

A receiver's internal memory holds such data as waypoints, track logs, routes, and uploadable digital maps (if the model supports them). The more memory the receiver has, the more data you can store in it. All the data that's been stored in the GPS receiver is retained when the device is turned off.

GPS receivers have different amounts of memory. Unlike personal computers, you can't add memory chips to a GPS unit to expand its internal memory.

Depending on the model, car navigation systems may use memory or a hard drive for internal storage.

External storage

An increasing number of GPS receivers aren't limited to internal memory for storage, using memory cards that can be plugged into the receiver to store data. External memory can be either

- **Manufacturer data cards**
- **Generic (and less expensive) memory card storage,** such as
  - MultiMediaCard (MMC; www.mmca.org)
  - Secure Digital (SD and microSD; www.sdcard.org)

User interfaces

All GPS models use a similar form of user interface where different types of information are shown on different pages. For example, you could scroll among pages that show you satellite status, a trip odometer, and your current course heading. For handheld models, each page typically has unique menu item commands associated with it for changing settings.

Each manufacturer has a hardware and software user interface for displaying and entering information. Sometimes the interface even varies between models.

Some GPS receivers have simple and advanced user-interface modes.

- **Simple mode:** This mode displays only commonly used commands and features.

  This is an excellent option for the novice who wants to use basic GPS receiver functions without being distracted or confused by the many other features.
- **Advanced mode:** This mode shows all commands and features.

Accessory software

Many handheld GPS receivers have built-in accessory programs that provide handy features, such as:

- Calendars with the best time to hunt and fish
- Sunrise, moonrise, sunset, and moonset tables
- Tide tables
- Calculators
- Geocaching databases
- Games

Accessory hardware

Optional built-in electronic gadgets decrease battery life when in use.

Electronic compass

All GPS receivers can tell you which direction you're heading — that is, as long as you're moving. The minute you stop, the receiver stops acting as a compass. To address this limitation, some GPS receivers incorporate an electronic compass that doesn't rely on the GPS satellites.

Like with an old-fashioned compass, you can stand still and see which direction your GPS receiver is pointing. The only difference is that you see a digital display on-screen instead of a floating needle.

On some GPS receivers, you need to hold the unit flat and level for the compass to work correctly. Other models have a three-axis compass that allows the receiver to be tilted.

Electronic compasses need to be calibrated whenever you change batteries. If your GPS unit has an electronic compass, follow your user guide's instructions to calibrate it. Usually, this requires being outside, holding the GPS unit flat and level, and slowly turning in a circle twice.

Personally, I think electronic compasses are an overrated feature. They consume batteries quicker, and I've heard more than one story of users forced to return their GPS unit because of malfunctioning compasses. Call me old school, but I'll stick with a traditional handheld, magnetic compass.

Altimeter

The elevation or altitude calculated by a GPS receiver from satellite data isn't very accurate. Because of this, some GPS units have *altimeters,* which provide the elevation, ascent/descent rates, change in elevation over distance or time, and the change of barometric pressure over time. Calibrated and used correctly, barometric altimeters can be accurate within 10 feet of the actual elevation. Knowing your altitude is useful if you have something to reference it to, such as a topographic map. Altimeters are useful for hiking or for when you're in the mountains.

The rough-and-ready rule is that if barometric pressure is falling, bad weather is on the way; if it's rising, clear weather is coming.

On GPS units with an electronic altimeter/barometer, calibrating the altimeter to ensure accuracy is important. To do so, visit a physical location with a known elevation and enter the elevation according to the directions in your user guide. Or, refer to Chapter 15 where I show you how to use Google Earth to determine a location's elevation. If you're relying on the altimeter/barometer for recreational use, I recommend calibrating it before you head out on a trip.

MP3 players

Some handheld and automotive GPS receivers let you jam to tunes while finding your way around.

Image viewers

With high-resolution screens, some newer GPS units can display photos and other graphics files.

Satellite radio receivers

Higher-end car navigation systems have satellite radio receivers that provide subscription-based traffic and weather information. I discuss this more in Chapter 6. (This feature, particularly receiving real-time weather maps and information, will eventually make its way to high-end handheld GPS units.)

The Future of GPS

Modern technology rapidly evolves, and the same holds true for GPS. Since consumer GPS receivers first became available in the mid-1990s, the market has grown tremendously because of cheaper receiver prices and new ways to use GPS. A peek into a crystal ball shows what the future may hold for GPS.

✓ **More accurate:** The United States has started planning the next generation of GPS, dubbed *GPS III.* Driving factors are better accuracy and reliability, improved resistance to signal jamming, and the looming European

Galileo system. Increasing the number of WAAS satellites in orbit is also planned. The target for GPS III is 2013.

✔ **Cheaper:** Prices will continue to decline as manufacturing costs decrease and marketplace competition increases.

✔ **Easier to use:** Simplified and less technical hardware and software user interfaces will be a priority as GPS receivers become more appliance-like to meet the needs of specialized uses and markets. For example, touch screens are becoming available on handheld outdoors receivers.

✔ **More integrated:** GPS receivers are being integrated into cars and trucks, cellphones, PDAs, Family Radio Service (FRS) and marine radios, watches, and other consumer electronic devices. Expect new products and services that take advantage of location-aware data. In the future, expect fighter-plane-style heads-up GPS displays to start appearing on car windshields.

✔ **Better maps:** Expect 3-D and photo-realistic maps to become common on both handheld and automotive GPS receivers.

✔ **Open source maps:** DeLorme's PN-20/40 GPS receivers can import various types of free government maps and charts available from the Internet. I expect other manufacturers to eventually, yet begrudgingly, move to this model.

✔ **SDKs and APIs:** I wouldn't be surprised to see GPS receivers with software developer kits and application programming interfaces just like cellphones and PDAs. A GPS unit that runs Linux or supports Java would open all sorts of interesting opportunities for third-party developers.

✔ **Less wired:** Most GPS receivers transfer data from personal computers through a cable. Wireless technologies, such as Bluetooth (`www.bluetooth.com`) and wireless USB, are well suited for fast and easy data transfers without using cables. The Dash Express (`www.dash.net`) car navigation system even features two-way Internet connectivity that provides a mesh network for drivers.

GPS monopoly

The U.S. has had a monopoly on satellite-based location systems over the years, but that will be changing in the future. At the end of 2007, the European Union agreed to fund the Galileo GPS program (Galileo was originally supposed to go online in 2008, but now is targeted for 2013). Russia is in the process of updating its GLONASS (Global Orbiting Navigation Satellite System), which is supposed to be globally operational in 2009. Not to be outdone, the Chinese are launching their global system, dubbed Beidou-2 or Compass. A number of countries don't like the power the U.S. exerts in being able to degrade or block GPS signals and want more of a level playing field. In the future, look for manufacturers to offer GPS receivers that work with multiple systems.

Chapter 4

Grasping Important GPS Concepts

To make the most of Global Positioning System, you should have a grasp of some important concepts — things like coordinate systems, datums, waypoints, routes, and tracks.

Even if you already own a GPS receiver and have used it for a while, this is still a good chapter to skim through because some of the basic concepts that I discuss end up being overlooked in user manuals. It's also difficult to keep all the concepts straight if you don't use your GPS receiver on a regular basis. Read on while I help clarify some important terms and concepts (like datums and waypoints and routes) so you'll be (pardon the pun) moving in the right direction.

Before you get started, I want to mention one more thing. The concepts in this chapter are important for outdoors, marine, and aviation GPS users. However, if you're exclusively using a car navigation system, you don't necessarily need a working knowledge of what I discuss. Coordinate systems, datums, waypoints, and routes are used, but kept conveniently hidden behind the curtains in most automotive GPS receivers.

Linking GPS, Maps, and Coordinate Systems

Some people think that after they have a GPS receiver, they really don't need a map, especially if the receiver has built-in mapping capabilities. This isn't necessarily true. GPS receivers are best used in conjunction with maps, whether those maps are paper or digital. Here are some of the reasons why:

- **Detail:** Most maps on handheld GPS receivers don't offer the detail of full-size paper or digital maps, especially topographic maps and nautical charts.

- **Size:** A GPS receiver's screen is pretty darn small, and it's just about impossible to get the big picture that a full-size map can give you.

- **Backup:** If you have a paper map with you and know how to use it, the map becomes an important backup if your GPS receiver's batteries fail or if you encounter poor satellite coverage. Gadget lovers might consider a paper map and magnetic compasses primitive, but they don't require batteries — and both are lightweight and cheap, to boot.

- **Complementary:** After you get back home or to the office, you might want to see where you've been on a map, based on the locations that you've stored in the GPS receiver. With a digital map, you can easily plot the exact route that you took or identify the places you visited.

Land navigation

Discovering how to effectively use paper maps and compasses for land navigation is both an art and a science, with complete books written on the subject. Here are two excellent, free online resources if you want to find out more:

- *Map and Compass for Firefighters* (NFES 2554) is a self-study course developed by the U.S. government for wildland firefighters. The course is available at `www.nwcg.gov/pms/training/map_comp.pdf`.

- *Map Reading and Land Navigation* (FM 3-25.26) is the U.S. Army field manual

on just what the title says, available at `www.globalsecurity.org/military/library/policy/army/fm/3-25-26/index.html`.

Another great way to become a pro land navigator is through the sport of orienteering. *Orienteering* involves using a map and compass to find control points (small flags) in the shortest amount of time as possible. Do a Web search for *orienteering* to find more about the sport and about how you can participate in clubs and events in your local area.

All maps and GPS receivers use *coordinate systems,* which are grids on maps that enable you to find locations on a map. Because GPS receivers are designed for use with maps, they support a number of coordinate systems that correspond to those commonly found on maps. (Chapter 2 has information on using map coordinate systems.) Thus, you can take a location that you recorded on your GPS receiver and precisely locate it on a map.

By default, your GPS receiver displays positions in latitude and longitude. However, you can change the settings to display locations in exotic-sounding coordinate systems, such as the Finnish KKJ27 grid, the Qatar grid, or the W Malaysian R grid. You probably won't need to switch to some of these obscure coordinate systems unless you're a world traveler, so you can stick with latitude and longitude or Universal Transverse Mercator (UTM), which are used pretty much everywhere. If you're fuzzy on what latitude, longitude, or UTM are, check out Chapter 2, which gives an overview of the two coordinate systems.

Suppose you have your GPS receiver set to latitude and longitude, and you record some locations. When you get home, you find that your paper map doesn't have latitude and longitude marks but uses only the UTM coordinate system. Not a problem. Look in your GPS receiver user manual for information on changing the coordinate system from latitude and longitude to UTM. This is a quick and easy way of converting data between coordinate systems. Alternatively, you can visit the Graphical Locator Home Page at Montana State University (www.esg.montana.edu/gl/index.html) to perform online conversions of latitude, longitude, and Township and Range.

Dealing with Datums

I mention in Chapter 2 that a *datum* is a frame of reference for mapping. Because the earth isn't flat, geographic coordinate systems use *ellipsoids* (a sphere that's not perfectly spherical, much like the shape of the Earth) to calculate positions on our third planet from the sun. A datum is the position of the ellipsoid relative to the center of the earth.

Unless you're a cartographer or geographer, that probably hurts just thinking about it. Sparing you a long and detailed technical description (which you can get in Chapter 2), here are two important things you need to know:

- **All maps have a datum.** Hundreds of different datums are in use all over the world. Most good maps used for navigation — and highway maps don't count — usually state which datum was used in making the map. Topographic maps usually list the data somewhere on the collar (white space edges).

How much latitude?

GPS units can display latitude and longitude in several different formats. Take the location at the top of Mount Bachelor (some great skiing in Central Oregon if you're ever in the neighborhood) and see how it can be expressed:

✔ Degrees, minutes, decimal seconds (D° M' S"): 43° 58' 46.94" N, 121° 41' 14.73" W

✔ Degrees, decimal minutes (D° M.M'): 43° 58.7824' N, 121° 41.2455' W

✔ Decimal degrees (D.D°): 43.9797069° N, 121.6874253° W

And just for fun, here are the UTM coordinates for that same location: 10T 0605273E, 48 70240N.

Ouch! Is this confusing or what? Although it doesn't seem like it, all these coordinates refer to the exact same location. Remember, just like converting locations from one coordinate system to another, you can also use your GPS receiver as a calculator to convert from different latitude and longitude formats. Suppose you have some coordinates in decimal degrees and need them in degrees, minutes, and seconds:

1. **Change the coordinate settings in your GPS receiver to decimal degrees.**

2. **Manually enter the coordinates as a waypoint.**

3. **Change the coordinate settings in your GPS receiver to degrees, minutes, and seconds.**

When you look at the coordinates of the waypoint that you entered, they're now displayed in degrees, minutes, and seconds. (*Note:* Because changing coordinate systems varies from model to model, check your user manual for specific instructions.)

✔ **You can set what datum your GPS receiver uses.** An example of a GPS map datum page is shown in Figure 4-1. You'll probably have to go through several layers of screen pages to get to it. For some reason GPS manufacturers seem to bury the page.

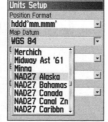

Figure 4-1: A GPS receiver map datum page.

The default datum for GPS receivers is WGS 84, more formally known as the World Geodetic System 1984. WGS 84 was adopted as a world standard and is derived from a datum called the North American Datum of 1983.

Many USGS topographic maps that you use for hiking are based on an earlier datum called the North American Datum of 1927, or NAD 27. This datum is divided into different geographic areas, so if you're in the United States — at least in the lower 48 states — use a version of NAD 27 that mentions the continental U.S.

So why is all this datum stuff so important? In the United States, if a position is saved in a GPS receiver by using the WGS 84 datum and the same coordinates are plotted on a map that uses the NAD 27 datum, the location can be off as much as 200 meters. That's more than a couple football fields off. The latitude and longitude coordinates will be identical, but the location is going to end up in two different spots.

The moral of the story is to make sure that the datums on your GPS receiver and maps are the same. Or, if you're with a group of people using GPS receivers, make sure that all your datums match.

Not having the map datum match the GPS receiver datum is one of the biggest errors that new users of GPS receivers make. I can't emphasize this point strongly enough: Make sure that the two match!

Setting Waypoints

A *waypoint* is GPS lingo for a location or point that you store in your GPS receiver. Some manufacturers also call them *marks* or *landmarks*. A waypoint consists of the following information:

- ✔ **Location:** The location of the waypoint in whichever coordinate system the GPS receiver is currently using. Some receivers also store the elevation of the location.

- ✔ **Name:** The name of the waypoint. You get to choose the name; the length varies between GPS receiver models from six characters on up.

- ✔ **Date and time:** The date and time the waypoint was created.

- ✔ **Optional icon or symbol:** An optional icon or symbol associated with the waypoint that appears on the GPS receiver's map page when the area around the waypoint is displayed. This could be a tent for a campground, a boat for a boat launch, or a fish for a favorite fishing spot. Many receivers also support entering a brief text description of a waypoint.

All GPS receivers can store waypoints, but the maximum number that you can save varies from model to model. Generally, as the price of a GPS receiver goes up, so does the number of waypoints that can be stored. Lower-end consumer GPS receivers store 250–500 waypoints; top-of-the-line models can store 1,000 or more.

Datum lessons learned

Here's a quick story from my Forest Service firefighting days that illustrates the importance of being aware of datums. A fire was reported in a mountainous area of eastern Oregon, and my partner and I helicopter-rappelled in to put out the fire. Because the fire was bigger than expected, we requested some smokejumpers to assist. They were down from Alaska, helping during the lower 48's fire season, and they all had new handheld GPS receivers, which were pretty state of the art back then (1998). The fire continued to grow, and we called in a small air tanker to stop the fire's spread. One of the Alaska jumpers pulled out his GPS unit, and we called in an exact set of coordinates for the pilot to hit. As the tanker approached, the pilot radioed us, asking whether we were sure that was where we wanted the retardant to go. Turns out that the jumper still had his GPS unit set with an Alaska map datum (that didn't match the local datum on the pilot's GPS receiver), and the coordinates that the Bureau of Land Management (BLM) jumper gave were on the other side of the ridge, nowhere near the fire. Fortunately, the pilot used his own initiative and dumped his load right where it needed to go.

Saving waypoints

The two types of waypoints you can enter and save to your GPS receiver are

- **Current location:** GPS receivers have a button on the case or an on-screen command for marking the current location as a waypoint. (Check your user manual for details.) After the waypoint is marked, the GPS receiver screen displays a waypoint information page where you can name the waypoint and associate an icon with it.

- **Known location:** If you know the coordinates of a location that you want to save as a waypoint, you can manually enter it in the GPS receiver. Most GPS receivers also allow you to mark wherever the cursor is on the map page as a waypoint. A known location could be a good fishing spot that a friend saved to his GPS receiver or perhaps a lake you want to visit that you got the coordinates of from a digital map. Again, check your user manual for directions on how to enter a waypoint manually. In Figure 4-2, you can see the Lowrance iFINDER waypoint page that displays after you press its ENT button.

 Always use meaningful names when you save a waypoint. GPS receivers typically assign a sequential number as the default waypoint name. Although numbers and cryptic codes might make sense when you enter them, I guarantee you that you probably won't remember what they mean a couple weeks down the road.

Figure 4-2:
A waypoint
page.

Some GPS units have a MOB function, which has nothing whatever to do with tommy guns or cement shoes. MOB stands for *Man Overboard* and was designed for boaters to use in case someone fell into the water. (Go figure.) After seeing or hearing the splash, the captain can press a button (which creates a waypoint appropriately named MOB) and then turn around and head back to the exact location of the unlucky sailor.

Although a GPS receiver is good for letting you know where you are, waypoints are important for assisting you to get somewhere you want to go. GPS receivers have a number of features that can help you navigate to a waypoint that you've entered.

Maybe you decide to go for a hike. Before you leave the trailhead, you save the location of the parking lot as a waypoint, naming it PRKLOT. *Note:* Some GPS receivers support waypoint names only in uppercase characters; others allow you to use both uppercase and lowercase characters.

Get in a habit of *always* creating a waypoint for your starting point on any outing.

While you hike down the trail, you hear the call of a rare ivory-billed woodpecker and head into the brush, intent on getting a sighting of the very elusive bird. After an hour of tromping around in the dense woods, you discover two things. First, the bird was just a robin; second, you're totally turned around and are somewhat lost. Fortunately, because you have a GPS receiver with you (and have read this book), you can easily find your way back to the car and be home in time for dinner. The following sections show you how.

Using the waypoint list

GPS receivers have an information page that lists all the waypoints that you've created and stored. (Again, check your user manual for information on how to access and use this list.) Waypoints can be listed by name or by those closest to your current location. By selecting a waypoint, no matter where you are, you can find

- ✔ **The distance to the waypoint,** such as a parking lot or trailhead.
- ✔ **The compass direction in degrees** that you need to get to the waypoint.

    In Figure 4-3, a GPS receiver screen provides information on how to get to a selected waypoint.

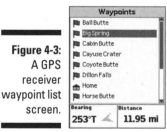

Figure 4-3: A GPS receiver waypoint list screen.

Any of the waypoints in the list can be deleted or edited. See your user manual for specific instructions.

If you roam around the Internet searching for information about GPS, you'll come across handy collections of waypoints that you can enter into your GPS receiver. There are all manner of waypoint lists, from fishing spots to pubs in England. If you're planning a vacation, consider doing a Web search to see whether any waypoints are associated with your destination. Then bring your GPS receiver along with you as a personal tour guide.

GPS receivers designed for outdoors always assume a straight line as the route between two points. That might be convenient for airplanes and boats, but it doesn't take into account cliffs, rivers, streams, fences, and other obstacles on land. GPS receivers designed for automobile navigation are a bit smarter, having built-in databases of road information that's used in suggesting and measuring appropriate routes from Point A to Point B.

Depending on the GPS receiver model, other waypoint-related information that you may be able to display includes

- ✔ **Travel time:** The amount of time it will take you to reach the waypoint based on your current speed.

- ✔ **Compass:** A picture of a compass that displays the waypoint direction heading.

- ✔ **Directional arrow:** An arrow that points in the correct direction that you should be heading.

- ✔ **Heads-up display:** A picture of a road that moves as you travel. If the road is centered onscreen, your destination is straight ahead. If the road veers to the right or the left, you need to correct your course so that the road is centered. A symbol associated with the waypoint grows larger as you get closer to it.

Some GPS receivers come with databases of cities, highways, airports, landmarks, and other geographic features. These are just waypoints stored in memory that you can't edit or delete to free up memory for new waypoints.

Most GPS receivers support mapping. At the very least, a GPS receiver has a simple plot display, a map page that shows waypoints, tracks (see the upcoming section, "Making Tracks"), and your current position. Many GPS receivers these days support maps that are more sophisticated; your waypoints and tracks appear along with roads, rivers, bodies of water, and whatever built-in features the map has. When the map page is displayed, you can zoom in, zoom out, and move around the map with an onscreen cursor that you control with buttons on the GPS receiver.

A map page can be orientated two ways: so the top of the screen always faces either north or the direction you're traveling. Orientating the screen to the north is probably the easiest if you're used to working with maps, which usually are orientated with their tops to the north.

Following Routes

A *route* is a course that you're currently traveling or plan to take. In GPS terms, a route is the course between one or more waypoints (see Figure 4-4). If multiple waypoints are in a route, the course between two waypoints is a *leg*. A single route can consist of a number of legs.

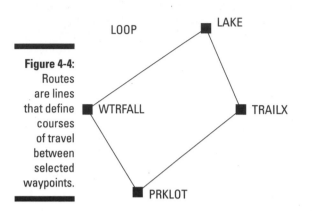

Figure 4-4:
Routes
are lines
that define
courses
of travel
between
selected
waypoints.

Suppose it's a beautiful day, and you go hiking, deciding to make a loop from a parking lot trailhead to a scenic waterfall, over to a lake for some lunch on a sandy beach, and finally cross-country until you reach a trail intersection that will take you back to your starting point at the parking lot. You've hiked in the area before; in fact, you've visited each of your planned destinations and marked them as waypoints in your GPS receiver. However, you've never hiked this particular loop before, so you decide to make a route called LOOP with the following legs that you've previously entered as waypoints:

PRKLOT to WTRFALL

WTRFALL to LAKE

LAKE to TRAILX

TRAILX to PRKLOT

After you create your route, the GPS receiver tells you how long each leg will be and the total distance of the route. When you *activate* the route (tell the GPS receiver you're ready to use it), the following information displays:

- ✔ **Direction:** The direction you need to travel in order to reach the next waypoint in the route
- ✔ **Distance:** How far away the next waypoint is
- ✔ **Duration:** How much time it's going to take to get there

After you reach a waypoint in the route, the GPS receiver automatically starts calculating the information for the next leg. This process continues until you reach your final destination.

Don't confuse a route with an *autoroute,* which applies to GPS receivers that can provide you with turn-by-turn street directions to a destination you're driving to. (I talk about routes and automotive GPS in Chapter 6.) AutoRoute is also the name of a Microsoft street mapping program for Europe.

Routes can be created ahead of time or entered while you're traveling. As with waypoints, after you create a route, you can delete or edit it, including removing or adding waypoints within legs.

Whenever you're using a route or navigating to a waypoint, you don't need to leave your handheld GPS receiver turned on all the time. Shut it off every now and then to conserve batteries. When you turn the GPS receiver back on, just select the waypoint or route that you were using, and the GPS receiver recalculates your present position and gives you updated information about how to reach your destination.

The number of routes and the number of waypoints that a route can consist of vary from one GPS receiver to another. Some inexpensive GPS receivers don't support routes, but a high-end consumer GPS unit might have up to 50 routes with up to 125 waypoints in each route.

Making Tracks

Remember the story of Hansel and Gretel, the kids who dropped breadcrumbs in the forest in order to find their way back home? Their story would've had a different ending if they had a GPS receiver that leaves electronic breadcrumbs (called *tracks* or *trails* depending on the manufacturer) while you travel. Every so often, the GPS receiver saves the coordinates of the current position to memory. This series of tracks is a *track log* or *track history.* (Because various GPS models handle tracks differently, check your user manual for specific details.)

Note these differences between tracks and waypoints:

- **Names and symbols:** Although tracks and waypoints are both location data points, tracks don't have names or symbols associated with them and can't be edited in the GPS receiver.

- **Autocreation:** Unlike waypoints — which you need to enter manually — tracks are created automatically whenever a GPS receiver is turned on (that is, if you have the track feature enabled).

If track logging is enabled, tracks are shown on the GPS receiver's map page while you move, like a trail of breadcrumbs. The GPS receiver constantly collects tracks while it's powered on, so you need to clear the current track log before you start a new trip. If you want, you can save a current track log.

If you turn your GPS receiver off or if you lose satellite reception, the GPS receiver stops recording tracks. When it's turned on again or good satellite coverage resumes, the GPS receiver continues recording tracks, but assumes that you traveled in a straight line between the last track location saved (before satellite reception was lost) and your current position.

To route or not to route

A fair number of handheld GPS receiver owners don't use routes and find them to be an over-rated feature. After all, after you reach your first destination, you can easily select the next location from the waypoint list and be on your way. Additionally, if you want to record where you've been, just using tracks is much easier. I discuss this in the "Making Tracks" section. However, here are some situations when you should consider using routes:

✔ If you travel to the same location on a regular basis (such as a guide leading clients on established trips).

✔ If you plan to share a route with other GPS receiver owners. Think of this as being a virtual tour guide. Routes can be down-loaded and then uploaded to other GPS receivers.

✔ If you have a mapping program, you can plan a trip ahead of time and create routes on your computer by simply clicking your mouse where you want to go. When you're finished, you can upload the route to your GPS receiver.

✔ If you're on a boat or an airplane.

Using routes is a personal preference. Try creating and using routes to see whether they meet your needs. If they don't, you can get by with waypoints and tracks.

Some GPS receivers allow you to set how often tracks are saved, either by time or distance intervals. For example, you could specify that a track be saved every minute or each time that you travel a tenth of a mile. Leaving the default, automatic setting for track collection should work for most occasions. However, if you're using your GPS receiver for specialized purposes (such as mapping a trail), you might want to experiment with different intervals to give you the level of detail you need.

When you reach your destination, your GPS receiver can optionally use the track log to help you navigate back to your starting point by using the track data to retrace your steps. Check your user manual for model specific instructions on how to do this.

Tracks are probably one of the most useful GPS receiver features if you're working with digital maps. From a number of free and commercial mapping programs, you can overlay your tracks on top of a map to see exactly where you've been. You can even edit the tracks (say, to correct for temporarily losing satellite reception). Figure 4-5 shows an example of track data collected during a trail run and then uploaded to a mapping program.

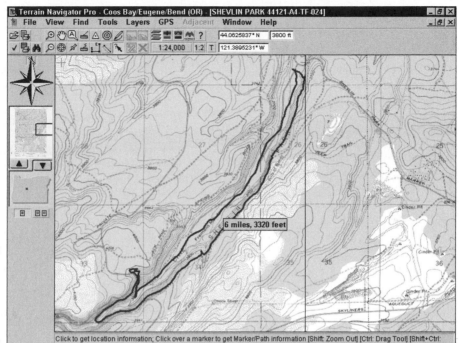

Terrain Navigator Pro - Coos Bay/Eugene/Bend (OR) - [SHEVLIN PARK 44121-A4-TF-024]

File   View   Find   Tools   Layers   GPS   Adjacent   Window   Help

44.0625837° N   3800 ft
1:24,000   1:2   T   121.3895231° W

6 miles, 3320 feet

Click to get location information; Click over a marker to get Marker/Path information [Shift: Zoom Out] [Ctrl: Drag Tool] [Shift+Ctrl:

Figure 4-5:
Track data
shows
where you
traveled.

Depending on the model, GPS receivers can store between 1,000 and 10,000 tracks and up to 10 track logs. If you exceed the maximum number of tracks, the GPS receiver will either stop collecting tracks or begin overwriting the oldest tracks that were collected first with the most current ones. (Some GPS receivers let you define what action to take.) The number of tracks collected over time depends completely on your activity, speed, GPS coverage, and the GPS receiver's track setting option. (I've found many handheld GPS receivers collect around 250 track points an hour on the default setting.)

Some GPS receivers reduce the number of tracks in saved track logs. For example, if you have 5,100 track points in the active track log, the number might be reduced to 750 track points when you save the log. This is done to save memory. You still have a general sense of where you've been, but you lose detail. If you need a high level of detail — such as when you're mapping a trail — always download the active track log to a computer before saving the track log to your GPS receiver.

You can download waypoints, routes, and tracks to your personal computer. The data can then be stored on your hard drive, used with digital mapping programs, or loaded into other GPS receivers. You can also upload waypoints, routes, and tracks that you create on your computer to your GPS receiver. See Chapter 10 for tips on how to interface your GPS receiver to a computer.

Can't we all get along?

You won't find any standards when it comes to GPS receiver waypoint, route, and track formats. Each GPS receiver manufacturer seems to have its own data format. To complicate things further, software companies that make mapping programs also use their own data formats. This can make exchanging GPS data between different receivers and software a very big challenge.

To help address this challenge, the folks at TopoGrafix (www.topografix.com/gpx.asp), a GPS and map software company, developed GPX. GPX stands for *GPS Exchange,* which is a lightweight, XML (eXtensible Markup Language) data format for exchanging waypoints, routes, and tracks between applications and Web services on the Internet. GPX built a lot of momentum and has been adopted by many software vendors and Web service providers.

Another way to exchange GPS data is with the free GPSBabel utility. This versatile program converts information created by one type of GPS receiver or software program into formats that can be read by others. GPSBabel is available for a number of different operating systems.

To get more information about GPSBabel and download the utility, visit www.gps babel.org.

Chapter 5

Selecting and Getting Started with a Handheld GPS Receiver

. . . .

. . . .

*T*his chapter is about selecting a handheld GPS receiver and getting started using it. Choosing a GPS receiver can be quite an overwhelming experience. If you look at all the handheld, portable GPS receivers on the market, you'll have somewhere around 50 GPS receivers to choose from (and that doesn't include discontinued models). That's a lot of choices. And that's only the beginning. After you purchase one, you still need to learn how to use it.

This chapter should take some of the confusion out of buying a GPS receiver and help you come up to speed using it. (*GPS For Dummies* isn't meant to replace your user manual, though, which you should refer to while you're reading the "Becoming Familiar with Your New GPS Receiver" section of this chapter.)

When it comes to selecting a GPS receiver, I won't recommend that you buy a particular brand or model or tell you which is best for hiking, *geocaching* (the best Easter egg hunt in the world; see Chapter 8), or other activities. Rather, I follow more of a Socratic method, in which I ask you a number of questions that should help you make a pretty good and informed purchasing decision.

Selecting a Handheld GPS Receiver

Handheld GPS receivers come in all sorts of shapes and sizes (as shown in Figure 5-1) and have a variety of features. Before you purchase a GPS receiver, you should spend some time kicking the proverbial tires. Don't rush out and buy a receiver based on one or two good Internet reviews without having a chance to hold that very GPS receiver in your hands to see how it works. Spend some time comparing different brands and models to determine which one works best for you. Because GPS units are sold in most sporting goods stores and many large retail chains, you shouldn't have to buy a receiver sight unseen.

Figure 5-1: Handheld GPS receivers, old and new, come in different shapes and sizes.

The top-selling manufacturers of consumer handheld GPS receivers in the United States are Garmin, Magellan (a part of Thales Navigation), Lowrance, and DeLorme. All these manufacturers have extensive Web sites that provide detailed information about their products. If you're in the market for a GPS receiver, definitely spend some time browsing through product literature. The Web site addresses for these manufacturers are

✔ **Garmin:** www.garmin.com

✔ **Magellan:** www.magellangps.com

✔ **Lowrance:** www.lowrance.com

✔ **DeLorme:** www.delorme.com

And don't just look at the marketing literature. Download the user manuals for the models you're interested in to better understand their features.

All GPS receiver manufacturers offer free Adobe Acrobat PDF versions of their product user manuals on their respective Web sites. If you're in the market for a GPS receiver, these are excellent resources for comparing features and seeing what the user interface is like because the manuals have instructions as well as screenshots.

Friends with GPS receivers are also a good source of information; ask to take their different brands and models out for a test drive.

Here are the two big questions that you should ask yourself before you begin your GPS receiver search:

✔ **What am I going to use it for?** Think about what activities you'll be doing with your GPS receiver: hiking, biking, fishing, sales trips, and so on. What will you expect your GPS receiver to do? Navigate streets or the wilderness, catalog favorite fishing spots, or find *geocaches* (hidden goodies from the popular electronic treasure hunting sport of geocaching)? When you get specific with your answers, you start to identify features that your GPS receiver should have to meet your needs.

✔ **How much do I want to spend?** How much money you have in your wallet or purse is obviously going to influence which models you end up considering. The more features a GPS receiver has, the more it's going to cost. So if you can figure out exactly what you're going to use the receiver for (see the preceding bullet) as well as which features you really need (versus those that are nice to have), you'll end up saving some money. Generally, figure on spending anywhere from a little under $100 to $500 for a handheld GPS receiver.

When loaded with the right maps and software, some handheld GPS models can function as an automotive navigation system, providing turn-by-turn street directions. When automotive GPS systems first came out, they were quite expensive. One inexpensive workaround was to use a handheld GPS unit that could be used for both driving and hiking. Nowadays, basic automotive GPS receivers are relatively inexpensive and include larger screens and more street-oriented features than you can get in a handheld. If you plan to use a GPS unit for outdoors and road navigation, I suggest you own both a handheld and automotive unit. (Cross-over automotive/outdoor models are available; I discuss the benefits and limitations of these in Chapter 6.)

For the most part, the cost of a GPS receiver really has nothing to do with accuracy. More expensive units tend to have advanced, sensitive chipsets though, which means satellite acquisition time is shorter and the receiver has an easier time locking onto signals. Figuring out how much you want to spend and what you want your GPS receiver to do narrows your options considerably, but you're likely still going to be faced with a number of choices. The next step is to pare down the list of candidates with some additional questions and things to consider, including:

- ✔ **Map display:** *Do you want to view detailed maps on your GPS receiver?* If so, you definitely need a *mapping model* — a GPS receiver that displays maps. See the upcoming section "To map or not to map."

   Function: *What activities do you plan to use your GPS receiver for?* I make recommendations for different features based on activities in the upcoming "Matching GPS receiver features to your activities" section.

- ✔ **Accessories:** *Does your budget include accessories, such as cases, cables, vehicle mounting brackets, and uploadable maps?*

- ✔ **Battery needs:** Consider the following questions:

   - *How many hours does the GPS receiver run on a set of batteries?* Remember two things: Different models (and their features) have different battery diets, and different battery types have varying life spans. (See the upcoming section "Battery basics" for the skinny on the different types of batteries and their life expectancy.)

   - *Will you need to carry spare batteries (always a good idea), and if so, how many?* I recommend always carrying at least one fresh set of spare batteries.

   - *Will you be using a cigarette lighter power adapter as an alternative to using batteries?*

- ✔ **Memory:** *How much memory does the GPS receiver have and is it expandable?* This is a critical question if you're interested in a GPS receiver that supports uploadable maps. (If you plan to store a lot of maps on your GPS receiver, consider a model that supports memory cards.) Visit the GPS receiver manufacturer's Web site to get an idea of how much memory maps can take. Better yet, check some of the Internet GPS information sources listed in Chapter 20, where users talk about how much memory they need for loading different types of maps.

- ✔ **Display screen:** Find out the following:

   - *How big is the screen and how well can you read it?* Make sure to consider visibility at night, in bright sunlight, and in poor weather conditions. The size of the screen is directly related to the overall size of the GPS receiver, so if you want a larger, more readable screen, expect a bigger GPS receiver to go with it.

- *Do you really need a color screen?* Most handheld GPS receivers feature color screens these days. A color screen makes reading maps easier because different colors are associated with map features. However, depending on your needs, color may be more of a preference than a requirement.

✔ **User interface:** *Does operating the GPS receiver make sense to you?* Sure, some learning is required to come up to speed, but using a GPS receiver should mostly be intuitive. Be sure to compare different brands and models because user interfaces are far from standardized.

✔ **External controls:** Look at different designs:

- *Are the buttons and controls on the GPS receiver easy to use?* Naturally, this is also related to the user interface.

- *Is a touch screen worth the extra price?*

- *Are the controls hard to operate while wearing gloves or mittens?*

✔ **Mounting hardware:** If you're going to use your receiver on a bike, motorcycle, or in a car, are brackets available for mounting it?

✔ **Weight and size:** *Do you want absolutely the smallest package you can get?* Note that there's only about a 7-ounce weight difference between the lightest and heaviest handheld GPS receivers. A bigger difference is the size of the receiver and its feel in your hand.

✔ **Computer interface:** *Do you plan to connect your GPS receiver to a computer to download and upload data?* If so, make sure that the receiver can interface with a computer to exchange data; I think this feature is a must so you can upgrade the GPS receiver's firmware. See Chapter 10 for more details on connecting GPS receivers to computers.

The beginning of 2008 saw brand-new handheld models introduced by several large GPS manufacturers. Touting many new features, the eagerly awaited units received a number of consumer complaints when they started shipping because of bugs in the firmware. The glitches were eventually fixed in subsequent software updates, but the events raise an important warning flag as manufacturers release new products quicker in an increasingly competitive marketplace. If you must have the latest and greatest GPS receiver, be prepared for the possibility of bugs when it first comes out; and be patient waiting for firmware upgrades to fix the problems. If you can wait a few months though, it's likely many glitches found by early adopters will be addressed, saving you some frustration.

To map or not to map

In terms of features, probably the biggest decision you need to make is whether to get a GPS receiver that displays detailed maps. Basic models typically don't support uploading topographic maps, marine charts, and street maps. Quite honestly, no matter what a salesperson might say, you don't absolutely need a GPS receiver that supports uploadable maps for activities such as hiking, geocaching, fishing, bird watching, kayaking, or other outdoor pursuits. Using waypoints (see Chapter 4) and tracks are all you need to navigate and successfully stay found. (Of course, you have a paper map and compass with you, and know how to use them, right?) Even though your GPS receiver doesn't display maps, if it can interface with a PC, you can still download information on where you've been and have it show up in a digital mapping program.

That said, a GPS receiver with a detailed map is pretty handy because it gives you a quick, big-picture view of where you're located in relation to other geographic and man-made features. And just the sight of a map, even though it's tiny, can be pretty reassuring at times even for a seasoned outdoors person.

Although I'm a firm believer that a mapping GPS receiver should never take the place of a paper map and compass, if your budget allows a mapping GPS receiver along with the digital maps to load with it, I say buy it. I personally use a mapping model for outdoor navigation on the land and water and treat the map feature as just another tool in my bag of "know where I am" tricks. In Chapter 11, find out more about the type of maps that you can use with GPS receivers.

GPS for athletes

Runners, bicyclists, and other outdoor athletes are turning to GPS receivers to help them track their workouts and become fitter and faster. A number of manufacturers offer GPS receivers that you wear on your wrist or mount on bike handlebars. (Figure 5-2 shows a Garmin Forerunner.) These devices are designed to cater specifically to the needs of athletes versus the average outdoors GPS user, and display speed, pace, distance traveled, calories expended, and other information. Some models even come with heart-rate monitors that let you view your heart rate in relation to your speed.

In addition to providing performance information while you workout, these specialized products also come with software that stores and analyzes information about your workout sessions. You can even overlay where you've been onto maps and aerial photos. (Check out the Motion Based Web site at www.motionbased.com for some ideas on what you can do with sports GPS data.)

Here is a list of manufacturers who offer GPS units designed for sport and fitness:

- ✔ **Garmin:** www.garmin.com
- ✔ **Polar:** www.polar.fi
- ✔ **Timex:** www.timex.com
- ✔ **Suunto:** www.suunto.com

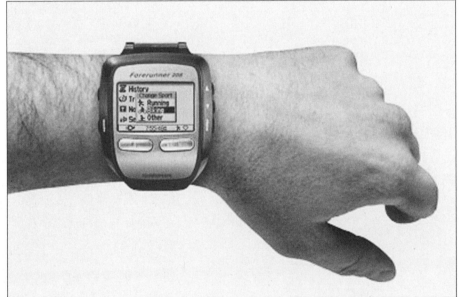

Figure 5-2:
A Garmin
Forerunner
wrist GPS
receiver.

Matching GPS receiver features to your activities

Most handheld GPS receivers are pretty versatile and can be used for a wide range of activities. However, some features make some GPS receivers more suited to certain activities than others.

Table 5-1 contains a list of activities in which people typically use GPS receivers as well as a list of features that could be useful for each activity. Just remember that these features aren't necessarily required and that a barebones GPS receiver will serve you equally well for basic navigation needs.

Table 5-1   Recommended GPS Receiver Features by Activity

Activity	Useful Features
Hiking, mountain biking, cross-country skiing	Altimeter/barometer Electronic compass Sunrise/sunset table Uploadable topographic maps
Geocaching	Electronic compass Geocache features Uploadable topographic maps
Hunting	Hunting calendar Sunrise/sunset table Uploadable topographic maps
Fishing	Fishing calendar Floats (it's buoyant) Saltwater tide table Sunrise/sunset table Waterproof
Boating (inland and offshore)	External power supply Floats NMEA* output (for autopilots) Saltwater tide table Sunrise/sunset table Uploadable nautical charts Waterproof
Canoeing, kayaking (inland and coastal)	Floats Saltwater tide table Sunrise/sunset table Uploadable nautical and topographic maps Waterproof
4 x 4-ing, motorcycling, ATV-ing	Electronic compass External power supply Uploadable topographic maps
Flying	External power supply Jeppesen database** WAAS
Mapping, data collection	Area calculation Differential GPS External antenna Helix versus patch antenna*** Large number of waypoints and tracks WAAS

Activity	Useful Features
Caving, scuba diving, visiting art museums	Sorry, you're out of luck! You need a clear view of the sky for a GPS receiver to work.

*National Marine Electronics Association.

**Jeppesen (www.jeppesen.com) is the leading U.S. supplier of pilot information. Its databases include information about airports, radio frequencies, and other flight-planning data.

***If there's tree cover, most users prefer the increased sensitivity GPS chip and a helix antenna. See Chapter 3 for more information about the differences between helix and patch antennas.

GPS receivers that are advertised as waterproof typically comply with the IEC (European Community Specification) Standard 529 IPX7. This standard states that a device can be immersed in up to 1 meter (a little over 3 feet) of water for up to 30 minutes before failing. Most handheld outdoors GPS receivers these days are IPX7 compliant. If yours isn't, and you're around water, I recommend you keep it in a waterproof bag.

If you're a pilot, you've probably figured out I don't include much information about GPS and flying in this book. If you want the lowdown on selecting and using GPS receivers for aviation, check out www.cockpitgps.com. John Bell, a commercial pilot, has written an excellent book — *Cockpit GPS* — that he makes available as a shareware PDF file. Highly recommended.

After you narrow down your choices to a couple of different GPS receivers, check out what people have to say about the models you're considering. Both www.amazon.com and http://reviews.cnet.com have lots of comments from consumers who have purchased popular GPS receivers. Perform a search for the manufacturer and model and read online feedback.

New GPS models are typically announced in January and released shortly before and during the annual Consumer Electronics Show. Check out Chapter 20 for Web sites that keep you updated on all the latest and greatest GPS models.

Becoming Familiar with Your New GPS Receiver

After considering all the options, making your list, checking it twice, and finding out which GPS receivers are naughty and nice, you've finally come to that blessed event where you're the proud owner of a GPS receiver. But before you rely on your new electronic gadget to guide you on a 100-mile wilderness trek or paddle up the Inside Passage, be sure to spend some time getting to know your GPS receiver.

To upgrade or not

So you already have a handheld GPS receiver, but some of those new models are catching your eye. What to do? Although I can't offer advice that pertains to an emotional want, I can offer suggestions based on practical needs. Here goes:

✔ **If you have an old, eight-channel model:** Definitely get a new unit. You'll be amazed at the improvements new technology has brought.

✔ **If you have a monochrome screen model:** Unless it's a specialized model, such as a sports-specific unit or a tiny Garmin Gecko (which I consider more of a secondary, accessory GPS receiver), it's time to upgrade to living color. Color is much easier on the eyes, especially if the model supports uploadable maps.

✔ **If you have a model from the early to mid-2000s:** This gets a bit tricky. The biggest advantages to post-2005 (or so) models are improved chipsets that lock onto satellites quicker and offer improved reception under trees and in canyons; memory card support for loading a whole lot of maps beyond what internal memory can store; and better screen technology. If your budget allows, a new model is worth considering, but definitely not a requirement.

✔ **If there's a feature on a new model that you absolutely have to have:** Who am I to spoil your fun, go for it. Just make sure your old GPS receiver goes to a good home instead of gathering dust in a closet. Outside of eBay or giving it to a friend or family member, schools, non-profit environmental groups, and search and rescue units are just a few of the organizations that might appreciate and use donated GPS technology.

A good place to start your GPS familiarization process is with the user manual. Many GPS receivers have a quick-start guide that gets you up and running in a matter of minutes. These guides are perfect for those impatient, got-to-have-it-now people; however, I suggest that you also take the time to read the full user manual (preferably in a location with satellite reception so you can try out different features). Otherwise, you could miss out on some important information contained in the full user manual.

In addition to the user manual, this section also helps you become familiar with your GPS receiver so you can get the most from it. Obviously, because so many GPS receiver models are on the market, don't expect to find detailed operating procedures for your specific model here: You need your user manual for that. What you can expect is basic information that applies to most handheld GPS receivers, including some things most user manuals don't mention.

Based on a number of years of search-and-rescue experience, I can list numerous occasions when hunters and hikers thought that their GPS receiver was some kind of magic talisman that would prevent them from getting lost. And quite often, when the search teams finally found them, they had no clue whatsoever how to use their GPS receiver properly. (There's a hint to bring the user manual with you, and if your PDA or cellphone supports it, store a PDF copy of the manual on it.) If you're going to rely on a GPS receiver for navigating outside of urban areas, take the time to find out how to use it so the friendly, local search-and-rescue people don't have to come looking for you. I'll step off my soapbox now, thank you.

Powering Your GPS Receiver

Before you can start using your GPS receiver, you obviously need to give it some power. For handheld GPS receivers, that usually means AA or AAA batteries. Manufacturers all give estimated battery lifetimes for their GPS receivers, but the actual number of hours a GPS receiver will run depends on how it's being used. For example, with the backlight on, battery life goes down because more power is consumed. In addition, what type of batteries you're using can also make a difference.

Battery basics

Although batteries may have the same size and shape, they definitely don't perform the same. Table 5-2 lists the pluses and minuses of some of the different types of batteries that you can use with your GPS receiver.

Table 5-2		Battery Comparison	
Type	*Rechargeable*	*Plus*	*Minus*
Alkaline	No	You can get these popular batteries just about anywhere. They're cheap (especially in quantity) and have a relatively long life.	Can't be reused.
Lithium	No	These batteries are lighter and work better under extreme cold conditions. They have the longest life and can be stored for up to ten years. These are different from the rechargeable Lithium-ion (Li-Ion) batteries found in laptops and cellphones. A few GPS receivers do have Li-Ion batteries.	Can be up to ten times more expensive than alkaline batteries; can't be recharged.

(continued)

Table 5-2 *(continued)*

Type	Rechargeable	Plus	Minus
Nickel cadmium (NiCad)	Yes	NiCads were the first generation rechargeable batteries, and they can be charged about 500 times.	Only have about one-third to one-quarter the life of an alkaline battery. Can develop a *memory* (can't be fully recharged). Have to buy a charger. Going the way of the dinosaur, replaced by NiMH batteries.
Nickel metal-hydride (NiMH)	Yes	NiMH batteries are much better than NiCads because they have a longer life (although not as long as alkalines), don't suffer from memory problems, can be recharged up to around 1,000 times, and are reusable.	Discharges when not in use (full recharge needed after several weeks to a month). Have to buy a charger.
NiMH Hybrid	Yes	Retains a significant charge when left unused. Can be charged more than 500 times. Highly recommended for GPS use. (Examples include Sanyo eneloop and Rayovac Hybrid.)	Higher cost than conventional NiMHs. Relatively new on the market. Requires a charger.
Rechargeable alkaline	Yes	Usually last about one-half to two-thirds as long as regular alkaline batteries and are rechargeable up to 100 times.	Cost roughly the same as NiMHs but aren't as popular. Charger required.

Battery tech 101

You can really get geeky with batteries and powering your GPS unit. If you get a charge out of electricity, here are some links to nitty-gritty information sources that cover voltage, milliamperes, and GPS drainage rates:

✔ **Battery drain for selected GPS receivers:** `www.gpsinformation.net/main/bat-5.txt`. This site offers the low-down on just how much juice different GPS receiver models consume.

✔ **NiMH Battery Shootout:** `http://candlepowerforums.com/vb/showthread.php?t=79302`. This site is oriented toward flashlights, but it has good and updated data on how various NiMH batteries perform. ***Note:*** NiMH batteries work best in high-drain, frequently used electronic devices like digital cameras.

They discharge relatively quickly when not in use and require recharging after several weeks to a month. For GPS units, I recommend the new generation of hybrid NiMH batteries, which don't discharge nearly as often.

✔ **Newsgroups:** `sci.geo.satellite nav`. Do a Google Groups search in this USENET newsgroup for batteries and be prepared to spend a couple of hours reading educational (and sometimes controversial) posts.

When you check these sources, you'll run into *mAh,* which means *milliampere-hours.* Most rechargeable batteries like NiMH have the mAh rating printed on their label. This rating is the battery capacity. Typically, the higher the mAh number, the longer the battery will last.

Expect AA alkaline batteries to cost from 50 cents to a dollar each (the more you buy in a single pack, the cheaper they are). NiMH AAs start at around $2 (with high capacity and hybrids costing more) while a charger can run you anywhere from $20–$50. (Some chargers come bundled with a set of two or more batteries.) When it comes to batteries and chargers, online prices are usually cheaper than full retail, so be sure to shop around.

If you're environmentally conscious or want to save some money over the long term, use standard or hybrid NiMH rechargeable batteries in your GPS receiver. Although a charger and pack of batteries obviously cost more than disposable alkalines, rechargeable batteries are a wise investment because they can be recharged hundreds of times before they end up in a landfill.

Even the best batteries can sometimes leak. If you're not going to be using your GPS receiver for a while, store it with the batteries removed.

Power to the people

After you select the type of batteries you're going to use, you should be aware of some other issues when it comes to powering GPS receivers:

✔ **Battery life gauges:** In the GPS receiver's setup information page, you can specify what type of battery you're using, such as alkaline or NiMH. The battery type setting helps the GPS receiver make an accurate guess regarding how long the battery will last. Remember that different battery types have different discharge rates. All GPS receivers also have an onscreen battery gauge that shows you how charged the batteries are. If you set the wrong type, the worst that will happen is that the gauge won't be accurate. See how to extend battery life with some GPS receiver models in the "Battery saver mode" sidebar.

Always check the battery level of your GPS receiver before you head out on a trip and remember to carry spare batteries. To tell which batteries are new or charged, put a rubber band around the good ones. By feeling around in your pack or pocket, you can instantly tell which ones are fresh. *Note:* Because conventional NiMH rechargeable batteries discharge faster than alkaline batteries when they're not in use, if you haven't used your GPS unit in a couple of months, don't be surprised if those rechargeable batteries are dead or don't have much life left. Because of this, I always carry alkaline or Lithium-ion batteries as spares.

✔ **Cigarette lighter adapters:** If you're primarily using your GPS receiver in a car or truck, you can save on battery costs by powering the GPS receiver with a cigarette lighter adapter. These handy devices run a GPS receiver from your car's electrical system. You can buy a generic version or one made for your model (sold by that GPS receiver manufacturer). Depending on the model, adapters cost between $20–$40, with the generic versions a bit cheaper than the manufacturer models.

✔ **Solar power:** If you have a GPS receiver that has a rechargeable Lithium-ion battery, you can do your part for the environment by using solar energy to charge it. Solio (www.solio.com) and Power Monkey (www.powermonkey-explorer.com) both have products that will work with GPS receivers that can be charged through USB ports. In addition to directly charging a Lithium-ion GPS unit, you can also hook up a solar charger to a USB battery charger, allowing you to keep your rechargeable AAs topped off. A solar charger is a great option for extended wilderness or expedition travel. Figure 5-3 shows a Solio charger attached to a Lowrance XOG receiver.

Cigarette lighter power adapters come with either straight or coiled cables. Although coiled cables are tidier, if your cigarette lighter isn't close to the dashboard, a coiled cable can pull your GPS receiver off the dashboard when it's not securely mounted. Adapters with straight cables don't have this problem; you can tidy up any slack in the cable with a plastic zip tie.

TECHNICAL STUFF

Battery saver mode

Some GPS receivers have a battery saver mode that can greatly extend the life of your batteries. (Check your user manual to see whether your model has this feature and how to turn it on.) Normally, a GPS receiver processes satellite data every second and determines your speed and location. Based on this information, the GPS receiver predicts where you should be the next time it gets satellite data. If the prediction is close to your actual position and battery saver mode is turned on, the GPS receiver will start receiving satellite signals every five seconds or so instead of every second. In addition, some of the internal electronics are turned off during this wait period. Because a reduced amount of power is needed, the battery life is extended.

The GPS receiver continues to access satellite data every 5 seconds until the predicted location isn't accurate anymore, at which time it switches back to receiving data every second, starting the process over again. (Some GPS receivers provide you with a number of choices of how often satellite data is received. The more seconds, the more battery efficient the receiver is.)

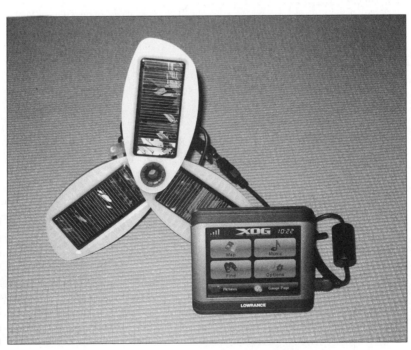

Figure 5-3: A Solio charger hooked up to a crossover GPS receiver.

Initializing Your GPS Receiver

Your GPS receiver now has power, so it's ready to go, right? Well, almost. After you put batteries in your GPS unit and turn it on for the first time, don't expect it to instantly display your location. A GPS receiver first needs to go through an initialization process before it can tell you where you are. The type of initialization and the amount of time it takes depends on what information the GPS receiver has previously received from the satellites and when.

The process is mostly all automatic, and you don't need to do much as your GPS receiver starts up and begins to acquire satellites. Your GPS user manual may contain model-specific initialization information.

To initialize a new GPS receiver, take it outside to a place with an unobstructed view of the sky (such as a large field or a park) and turn on the power. (You did install the batteries first, right?). After the start-up screen displays, the receiver will begin trying to acquire satellites.

It can take anywhere from a few minutes to 30 minutes for the GPS receiver to gather enough satellite data to get a position fix for the first time (usually more toward the 5 minutes or less end of the scale). Don't worry; your GPS receiver isn't going to be this slow all the time. After the GPS receiver is first initialized, it usually only takes 15–45 seconds under good conditions to lock on to the satellites when you turn it on. Newer GPS receivers with more sensitive chipsets can lock onto satellites even faster.

To speed up the location fix the first time or when the GPS receiver has been moved hundreds of miles since it was last turned on, many GPS receivers have an option where you move a cursor on an onscreen map of the United States or world to show your general location. Providing a rough location helps the GPS receiver narrow its search for satellites that are visible from your present location, speeding up the initialization process.

Manufacturers often use the terms *cold start* and *warm start* to describe different GPS receiver start-up states and times. Unfortunately, their definitions of these terms can be different, which makes comparative information about start times not very useful. Manufacturers have gotten clever when it comes to start-up times by displaying splash screens and dialog boxes a user has to interact with while the unit locks on to satellites. This makes the satellite-acquisition time seem faster by the time the main screens display.

It takes longer for a GPS receiver to initially determine its position when it's moving. If you have the option, keep your receiver stationary until it locks on to at least four satellites.

GPS receiver initialization nitty-gritty

You really don't need to know this technical information to operate your GPS receiver, but to start acquiring satellites to get an accurate location fix, a GPS receiver needs the following satellite data:

✔ A current almanac (rough positions of all the satellites in orbit)

✔ The GPS receiver's current location

✔ The current date and time

✔ *Ephemeris* data (precise position of individual satellites)

If some or all the data is missing or out-of-date, the GPS receiver needs updated information from the satellites before it can accurately fix a current position. The types of data that are out-of-date or missing determine how long the GPS receiver takes to initialize. If the GPS receiver is brand new, out of the box, several hundred miles away from where it was last used, or has been stored for a prolonged period, initialization will take longer.

Most GPS receivers show a satellite status page while the receiver is acquiring satellites; Figure 5-4 shows an example status page. This page typically consists of two circles that represent a dome of sky above your head. The outer circle is the horizon, the inner circle is 45 degrees above the horizon, and the center of the inner circle is directly overhead. The N on the page represents north.

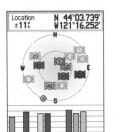

Figure 5-4: A GPS receiver's satellite status page.

Based on the almanac information, the GPS receiver shows the position of satellites within the circles, representing them with numbers. When a signal from a satellite is acquired, the number is highlighted or bolded. WAAS satellites are also shown if available.

Underneath the circles are a series of bar graphs that represent signal strength. (Depending on the model, the bar graphs may have numbers that correspond to the satellites that the GPS receiver has located.) The more a bar is filled in, the better the GPS receiver is receiving signals from that particular satellite. Depending on the model, signal strength may also be shown in color.

Try moving your GPS receiver to watch the satellite signal strength change. If signals are weak or you get a message about poor satellite coverage, move to another location and change the position of the receiver to better align it with the satellites that are shown onscreen. If you're successful, you'll see new satellites acquired, the signal strength increase, or both. The more satellites you acquire and the stronger the signals, the more accurate your reported location.

Holding the GPS receiver properly optimizes signal reception. If your GPS receiver has a patch antenna, hold it face up, parallel to the ground. If your GPS receiver has a quad helix antenna, hold it straight up so that the top of the receiver is pointing toward the sky. Chapter 3 covers the differences between patch and quad helix antennas.

After the GPS receiver gets enough information from the satellites to fix your location, the screen typically displays an Estimated Position Error (EPE) number. Based on the satellite data received, this is the estimated error for the current position. The smaller the number displayed, which will be in feet or meters, the more accurate your position.

Estimated Position Error (EPE) is a bit confusing. If you see an EPE of 20 feet, it doesn't mean that you're within 20 feet of the actual coordinates. You're actually within up to two times (or even more) the distance of the EPE from the actual location. For example, if you have an EPE of 50, your location could be 1–100 feet from the actual coordinates. EPE is not a maximum distance away from the actual location; it's only a measurement estimate based on available satellite data. To complicate things even further, different GPS receiver manufacturers use different formulas for determining EPE, so if you set three different GPS receiver brands next to each other, they all display different EPE numbers. Some manufacturers are conservative with their numbers, and others are optimistic. Don't get too caught up with EPE numbers; just treat them as ballpark estimates — and remember, the smaller the number, the better.

Trimble planning software

During certain parts of the day, you might have better satellite coverage than at other times because of the number of satellites that are in view and the position of a single satellite relative to the GPS receiver and other satellites in the constellation.

Trimble Navigation (www.trimble.com), one of the largest manufacturers of commercial and professional GPS receivers, has a free Windows program called Planning, shown in the figure here. Planning is designed for surveyors who need to know the best time to use their GPS surveying instruments. By entering the latitude and longitude coordinates of a location and the date, Planning gives you information on

✔ **Dilution of Precision (DOP):** DOP describes how accurate a reported GPS position is. The smaller the DOP number, the higher the accuracy.

✔ **Satellites:** You can see how many satellites are in view if the sky is unobstructed, the optimum times of satellite visibility, and the satellite orbit paths.

You don't need to be a surveyor to use this information. Knowing optimal GPS times is useful for all sorts of outdoor activities. For example, if you're serious about geocaching, you can select the best time of day to place a cache when your GPS receiver gives you the most accuracy. If your GPS receiver supports it, you can also use its averaging feature. This samples the recorded coordinates for a location and uses an average as the position.

Planning is easy to use and works for any location in the world with all the GPS satellite information presented in graphs or lists.

To download Planning, go to `www.trimble.com/planningsoftware_ts.asp?Nav=Collection-8425`.

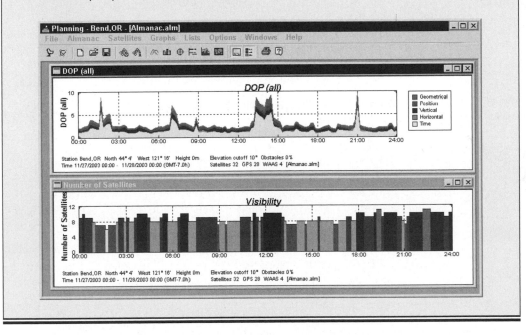

Changing Receiver Settings

After you initialize your GPS receiver for the first time, you need to change a few of the receiver's default system settings. You only need to do this once, and a few GPS receivers will prompt you to make some of these changes as part of the initialization process. These changes are mostly to customize settings based on your location and needs. Check your user manual for specific information on how to change the system settings described below. An example of a GPS receiver system settings page is shown in Figure 5-5.

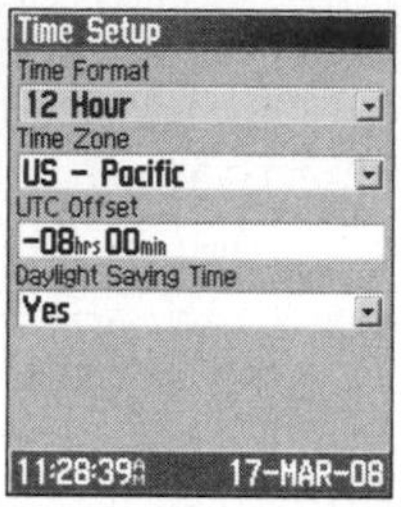

Figure 5-5:
A GPS
receiver
system
settings
page show-
ing time
options.

```
Time Setup
Time Format
12 Hour              ▼
Time Zone
US – Pacific         ▼
UTC Offset
–08hrs 00min
Daylight Saving Time
Yes                  ▼

11:28:39ª    17-MAR-08
```

Although GPS receivers have a number of system settings that you can change, here are some of the important settings you'll want to initially adjust:

✔ **Time:** Your GPS receiver gets very precise time data from atomic clocks aboard the satellites, but it's up to you how the time will be displayed. You need to specify

   • Whether to use 24-hour (military) or 12-hour (a.m. and p.m.) time

   • Whether Daylight Savings Time is automatically turned on and off

   • What your time zone is (or your offset from UTC)

   Your GPS receiver gets time data from the satellites in the UTC format. UTC stands for Coordinated Universal Time (no, the acronym doesn't match the meaning), an international time standard. UTC is a time scale kept by laboratories around the world, using highly precise atomic clocks. The International Bureau of Weights and Measures uses time data collected from the labs to establish UTC, which is accurate to approximately 1 *nanosecond* (about a billionth of a second) per day. In 1986, UTC replaced Greenwich Mean Time (GMT) as the world time standard. The Greenwich meridian (prime meridian, or zero degrees longitude) is the starting point of every time zone in the world. GMT is the mean time that the Earth takes to rotate from noon to the following noon. These observations have been kept since 1884 at the Royal Observatory in Greenwich, England. In hours, minutes, and seconds, UTC and GMT always have the same values.

✔ **Units of measure:** Your GPS receiver can display distance information in statute (such as feet and miles), nautical (knots), or metric (meters and kilometers) formats. The default setting for GPS units sold in the United States is statute, so unless you're boating or want to use the more logical metric system, leave the setting as is.

✔ **Coordinate system:** By default, your GPS receiver displays positions in latitude and longitude. If you want to use location coordinates in a different format, now's the time to change the setting.

- **Datum:** The default datum for all GPS receivers is WGS 84. Unless you're planning to use your receiver with maps that have a different datum, leave the default setting. (See Chapter 4 for more information on types of datums and their impact on GPS position coordinates.)

- **Battery type:** The default battery setting on most GPS receivers is alkaline. If you're using another type of battery, select the correct type. The battery type setting doesn't affect the GPS receiver's operation; it only ensures that the battery life is correctly displayed on the screen because different types of batteries have different power characteristics.

- **Language:** Most GPS receivers are multilingual, so if you'd rather view the user interface in a language other than English, it's as simple as selecting a different language from a menu.

Using Your GPS Receiver

Finally! After you initialize your GPS receiver and change some of the system settings, it's time to use it. Start with

- Going through the GPS receiver's different onscreen pages and seeing what information is displayed.

- Walking around and watching what happens to the numbers and your position on the GPS receiver's mapping and trip pages. (Do this outside, of course.)

- Pressing buttons and seeing what happens. You may want to have your user manual nearby in case you get lost between information screens.

GPS receivers are pretty robust, and you're not going to hurt your new purchase by being curious.

GPS receiver screens can be scratched relatively easily. Investing in a carrying case keeps the screen scratch-free; many cases have a clear plastic face that allows you to use and view the GPS unit without taking it out of the case. Another way to keep the screen from being scratched is to buy thin, clear plastic sheets used to protect PDA screens, cut them to shape, and place the sheet on top of the GPS receiver screen.

The following are some simple exercises you can try that will help you become familiar with your GPS receiver. When you first start using your GPS receiver, take the user manual with you. If you forget how to do something or have a question, the manual will be right there for reference.

Coming home

As Dorothy says in *The Wizard of Oz,* "There's no place like home." Here's an exercise that will always help you find your way back (or at least know how far away home is). Clicking your heels three times is optional.

1. **Take your GPS receiver outside where you live and create a waypoint for the spot where you're standing.**

   See Chapter 4 for the skinny on creating waypoints.

2. **Name the waypoint HOME.**

3. **Turn the GPS receiver off and go for a walk.**

   How far is your call, but at least travel far enough that you can't see your starting point.

4. **When you're ready to head back home, turn the GPS receiver back on and use it to navigate back to the HOME waypoint.**

   Be sure to move through the different onscreen pages to watch the direction and distance change while you head back home.

After you enter the HOME waypoint, no matter where you are, if you have your GPS receiver with you, you can always tell exactly how far away and what direction home is. Remember, this is in a straight line as the crow flies unless you have a GPS receiver that supports street navigation.

How far, how fast?

Your GPS receiver also contains a very accurate trip computer that displays information about distance, speed, and time. After you read your user manual on how to reset and start the trip computer, here are some ideas for getting familiar with how it works:

✔ **When exercising:** When you walk, run, jog, bike, or whatever, take your GPS receiver with you on your favorite course to see just how far you go. At the end, check your average and maximum speed.

✔ **When going to school:** If you have children and they walk to school, go with them on their route to see exactly how far it is. (And tell them when you were young you used to have to walk at least ten times that distance . . . in the snow . . . uphill both ways.)

✔ **When doing yard work:** The next time you cut the grass, take your GPS receiver with you and see just how far you push your lawn mower.

You don't need to keep your GPS receiver in your hand for it to work. Usually, it will receive satellite signals just fine tucked in a jacket pocket or in the top of a pack.

Finding your ancestors

Many people are into genealogy these days, and your GPS receiver can be a helpful tool in tracking your ancestors. When you visit a cemetery looking for long-lost kin, bring your GPS receiver with you to record the exact locations of tombstones and grave plots. You can pass the latitude and longitude on to other relatives doing genealogical research. The coordinates can be extremely useful for someone locating a small out-of-the-way cemetery in the countryside, or a relative buried in a cemetery with thousands of plots.

Simulating navigation

Some GPS receivers have a simulator or demonstration mode. This is probably one of the most overlooked (but coolest) features on a GPS receiver. The simulator mode acts as if the receiver is actually acquiring GPS satellite information. You select a speed and a direction, and the GPS receiver pretends you're moving. Because the receiver isn't relying on acquiring satellite data, you can comfortably sit inside the house in your favorite chair, getting familiar with your new purchase.

Depending on where you live or work, how many windows you have, and your view of the sky, your GPS receiver might (might) work indoors (or at least close to windows). Although you're limited to what you can do with a GPS receiver indoors, it's fun to see just how much GPS coverage you can get by walking around inside a building.

Homework assignment

Here's a little ungraded homework assignment for you to see how well you know your GPS receiver. And yes, you can peek in the user manual.

- ✔ Change the coordinate system setting from degrees and decimal minutes to degrees, minutes, and seconds (or vice versa).
- ✔ Change the map display from north always being at the top to north being oriented to the direction of travel.
- ✔ Change the datum from WGS-84 to NAD27, and then change it back.

✔ Write down the number of satellites your GPS is receiving signals from and the EPE.

✔ Write down what time the sun sets today (if your unit supports this feature).

✔ Write down your current elevation.

That's the end of the assignment. How did you do?

And more . . .

You can do lots more with a GPS receiver besides basic navigation. Think outside the box. Some examples include

✔ Take digital pictures of cool places and record their coordinates with your GPS receiver. You can post them on a Web site or e-mail them to friends. There's even free software available for embedding GPS coordinates into digital photographs. This is *geotagging,* and I talk more about it in Chapter 7.

✔ If you have a smaller GPS receiver, securely attach it to your dog's collar and track where Fido goes for the day. (You can also find commercial GPS pet locator products on the market, which I discuss in Chapter 7.)

✔ Play a game. Although lots of people have heard of geocaching (if you haven't, check out Chapter 8), there's also GeoPoker, GeoGolf, and GeoDashing to name a few. Visit www.gpsgames.org to see whether any of the activities strike your fancy.

✔ Use your track log to create art. Some GPS users express themselves as artists by using their GPS receiver to record their movements as they walk around trying to create shapes or pictures. (Don't believe me? Check out www.gpsdrawing.com.)

Your goal should be to become confident using your GPS receiver and to have fun in the process.

Virtual GPS

Lowrance (www.lowrance.com) has a unique series of simulators that run on your computer for some of its sonar and GPS products. Just download and run a program from the Internet, and a lifelike, reduced-size replica of the product appears onscreen. The GPS receiver simulators don't actually track satellites, but other than that work just like the real thing. Use your mouse to click buttons, and the keyboard arrow and Enter keys to select menu items and change settings.

Although Lowrance designed the simulators as a way for potential customers to become familiar with their products (and for owners to practice with them), GPS receiver simulators are an excellent way to find out about concepts and functions that are common to all GPS receivers, no matter what the brand or model. The user interface and features vary between brands and models, but key GPS receiver concepts such as datums, coordinate systems, waypoints, routes, and tracks remain the same. The Lowrance simulator, shown in the figure here, lets you get up to speed on basic GPS receiver operations without even owning a receiver; plus, it's fun to play with. (If you download a simulator, get a copy of the real product's user manual so you can understand and try all the features. These can be downloaded from the Lowrance site as well.)

To download Lowrance simulators, go to www.lowrance.com/software/pcsoftware/demos.asp.

Chapter 6

Automotive GPS Receivers

. . . .

In This Chapter

▶ Reviewing automotive GPS receiver features

▶ Understanding different types of automotive GPS receivers

▶ Selecting an automotive GPS unit

▶ Using your automotive GPS receiver

. . . .

*T*his chapter is about selecting an automotive GPS receiver and getting started using it. These GPS units are specifically designed for cars, trucks, and RVs that travel on paved roads, streets, and highways. Although automotive GPS receivers use the same technology as handheld GPS models for keeping you found, they are significantly different enough that I want to devote an entire chapter to them.

When automotive GPS units first appeared on the market several years ago, they were quite expensive and only a handful of models were available. Since then, things have changed considerably. Prices have dropped and retailer shelves are filled with a bewildering number of brands and models.

This chapter should take some of the mystery out of buying a car GPS receiver and help you come up to speed using it. (Just remember though, this book isn't meant to replace your user manual.)

I start by listing some of the common features you find in automotive GPS units and then talk about the different types of models on the market. With that in mind, I then ask you some questions that will help you make an informed decision on which unit is best for you. Finally, I wrap up the chapter with some useful tips on using your new GPS receiver.

Automotive GPS Receiver Features

All automotive GPS receivers share a number of similar features and characteristics. And just like their handheld cousins, the more features they have, the more the price tag tends to go up.

Before you start shopping for a car GPS unit, it's useful to know what features you'll encounter. Let's check them out.

Street map

You probably already guessed that an automotive GPS receiver's main job is to serve as an electronic street map. While you drive, your current location displays on an onscreen map, which constantly moves, matching the speed and direction you're traveling. An example map is shown in Figure 6-1.

Figure 6-1:
A typical automotive GPS receiver street map.

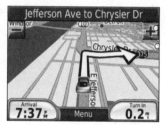

Maps that come with an automotive GPS unit typically cover a large area, such as all of the United States (usually Canada too), so you don't need to purchase a map for each state you travel in.

The street map on the screen isn't like a simple paper map. A number of enhancements make the map much more useful for driving and street navigation, such as the following:

✔ The map can be displayed in a 2-D overhead view or simulated 3-D perspective.

✔ A night mode is available that changes screen colors, making the map more readable in the dark.

✔ The level of map detail and scale can be controlled by zooming in and out.

✔ Icons appear on the map showing the location of restaurants, gas stations, and other services (see the upcoming "POIs" section).

✔ Driving information, such as current speed, time, and compass direction, can selectively be displayed on the map for reference.

Routing directions

In addition to showing you where you currently are, an automotive GPS receiver also gives you directions on how to get somewhere you want to go.

Unlike handheld GPS units that use latitude and longitude to reference locations, automotive GPS receivers use much more familiar and friendly street addresses (behind the scenes, street addresses are converted into latitude and longitude for mapping purposes through a bit of technical magic called *geocoding*).

You enter a street address and city or ZIP code, and the GPS unit calculates a route to get there from your current location (either the shortest or the fastest route, which might not necessarily be the same). Unlike a handheld GPS receiver that plots a straight, as-the-crow-flies line between two points, an automotive unit consults road data stored in memory and provides you with turn-by-turn street directions to a chosen destination.

Many models have options that allow you to further customize a route, such as "avoid toll roads" or "stay on highways," go through a selected city, or specify multiple destinations. Check your user manual for more details.

Automotive GPS receivers tend to be pretty smart, so if you deviate from a route for some reason, the unit will automatically recalculate a new route and provide directions from your current location.

Some models detect if you're moving too slowly (or not at all) on a freeway and offer an alternative route to detour you around a traffic jam. You can specify the distance the detour will take you around the problem, such as, "Detour me on another road one mile past where I am now." (This is different from real-time traffic information services, which I talk about soon.)

While you travel, directions are displayed on the screen three different ways:

- ✔ A colored route path is overlaid on the map, showing which streets and roads you need to take to reach your destination.

- ✔ When you approach a turn, a small picture of a right arrow or left arrow is shown, indicating the direction you should turn. (Other prompts with images are used, such as an icon that shows getting on a freeway on-ramp.)

- ✔ An optional list that shows turn-by-turn directions, with the distance between each turn, and how far you are from the next suggested turn (which is typically shown at the top of the list).

In addition to visual cues, there are also sound and voice prompts. These include

- ✔ **Voice prompts:** A spoken voice provides simple directions as you approach a turn, such as, "In 500 yards, turn right." When you're at the turn, the GPS unit announces, "Turn right." Because most directions consist of distances and right or left turns, it's easy to build a very limited vocabulary of voice prompts into a GPS unit.

✔ **Text-to-speech:** A step above voice prompts, and typically found in mid- to higher-end GPS models, is text-to-speech. Instead of having a limited vocabulary of canned directions, the GPS unit has a built-in speech synthesizer that pronounces street names in its routing database. For example, "In 500 yards, turn right on exit 12" or "Turn left on Davis Boulevard."

Sounds and speech are played through the GPS unit's internal speaker. Some models also have features for broadcasting directions through your car's stereo system.

If your GPS receiver can broadcast over your radio, find an unused FM frequency your GPS unit can use by visiting www.radio-locator.com/cgi-bin/vacant.

Speech synthesizers use a standard set of pronunciation rules. Don't be surprised if you sometimes hear street names that sound a little odd because they're pronounced incorrectly. Most speech synthesizers also support giving directions in languages other than English, such as Spanish and French.

Streets can be known by multiple names (for example a main drag that runs through my city is variably known as Highway 97 or 3rd Street or Division Street or U.S. 97). The GPS receiver will use only one name, which its database believes is the most commonly used. This can cause some confusion if the GPS unit tells you to turn on a street, and the directions don't happen to match the name on the street sign.

In addition to spoken directions, automotive GPS receivers may have tones or bells to alert you before a turn. Handheld GPS receivers that use street navigation software reply exclusively on tone prompts.

POIs

POI stands for Point (or Points) of Interest. A POI can be a restaurant, gas station, bank, park, rest stop, or any number of other places on the road you may want or need to visit. Automotive GPS receivers have a large database (in the millions) of POIs associated with the built-in electronic street maps. POIs appear as icons on the map, with different icons representing different types of services (for example, a fork and spoon might symbolize a restaurant).

If POIs aren't appearing on your map, try zooming in. Most automotive GPS receivers have settings of at which zoom level to display POIs to avoid cluttering up the screen. You can also set options to show only certain types of POIs on the map.

When you select a POI, its name displays next to the icon. Most GPS units can also display more detailed information about a POI, such as an address and phone number.

Some manufacturers offer free software that allows you to create custom POIs and transfer them to your GPS receiver. Check your user manual or the manufacturer's Web site to see if your model supports this feature.

Many automotive GPS receivers support the ability to search the POI database. For example, if you're on a trip and starting to get hungry, your GPS unit can display a list of nearby restaurants based on your current location. If you're in the mood for Chinese, you can even request the unit show only Asian restaurants. After you find one that looks good, press a button on the screen, and your GPS unit will calculate a route and give you directions for getting there.

Some GPS models have more POIs than others do (and some manufacturers add premium information from AAA and Zagat). If you find yourself relying on this feature, the more POIs a GPS receiver has, the better.

The companies that compile POI data do their best to keep comprehensive and up-to-date lists. However, don't expect every business and service to appear in a POI database or all the information to be 100 percent accurate. The constant cycle of new businesses starting and old businesses closing makes it challenging to stay current. To learn more about updating your POI database, see the "Updating maps and firmware" section at the end of this chapter.

Address books

Because automotive GPS products are based on street addresses, it makes sense to incorporate address books into the units. An address book stores a location's name (which could be Grandma's house, Taco Stand, Jewell's By the Sea Inn, or any descriptive name), its associated street address (including city and state), a phone number, and any other pertinent information.

After you save a location to the address book (by keying in the information on the touch screen), you simply select the location and your GPS unit plots a route to it. No need to type in the exact address each time.

Trip Log

All automotive GPS units have a trip log page that tracks time and distance you've spent on the road. Think of it as a high-tech version of your car's odometer.

Screens

The majority of automotive GPS receivers have touch screens. This means you enter commands and interact with the unit by touching different parts of the screen. Just like a computer program, GPS receivers have windows or pages with various types of information and commands. This is an easy to learn and practical way of controlling the GPS receiver, although if you have large hands and thick fingers, it sometimes can be a bit challenging.

Invariably, your GPS receiver screen is going to get fingerprint smudges. Keep it clean with a soft, micro-fiber cloth or a cleaning kit designed for laptop computer monitors. Using household cleaning products may damage the screen.

A readily apparent feature that sets apart different automotive GPS receivers is the screen size. You'll find a number of different screen sizes, from compact to large, rectangular shaped to wide screen (as shown in Figure 6-2). The general rule of thumb is the larger the screen, the more the product will cost.

Figure 6-2: Standard and wide-screen automotive GPS units.

Preferences

All automotive GPS receivers have one or more preferences pages that display settings and options, such as speaker volume, screen brightness, amount of map detail, and other parameters.

Most people rely on the default preferences of their GPS unit, but learning about the various settings allows you to customize the receiver to make it more usable. Check your user manual to see what you can do!

Entertainment features

As competition in the automotive GPS receiver heats up, and manufacturers look for new ways to distinguish their products, non-navigation features are starting to become common. Some common entertainment features include

✔ **MP3 player:** Plays songs on an inserted memory card over the GPS receiver's speaker or through a headphone jack. Some higher-end models send the tunes to your car's stereo system by broadcasting the music through unused FM radio frequencies.

✔ **Image viewer:** Displays JPG and other graphics files stored on an inserted memory card.

✔ **Games:** Some manufacturers are incorporating games into their automotive products.

In the spring of 2008, a few GPS units destined for the Asian market started to show up that supported receiving digital TV broadcasts. It's just a matter of time before this feature becomes available on this side of the Pacific. As if normal GPS receivers, cellphones, and stereos weren't enough of a distraction while driving!

Most automotive GPS units that have USB ports and memory cards serve as UMS mass storage devices when connected to a PC. That means they show up as an external drive (such as E:\) just like a USB thumb drive or portable hard drive.

Wireless

Wireless connectively is another feature you'll find in some automotive GPS products. The most common feature is Bluetooth support that allows you to connect to a Bluetooth compatible phone and control it with the GPS unit.

A more cutting-edge, wireless implementation, that could provide a glimpse into the future, is Internet connectivity. In 2008, start-up Dash Navigation (www.dash.net) introduced a new automotive GPS product that sends and receives information through WiFi and cellular data networks. A Dash GPS unit collects information about where you're driving and how fast you're going. This real-time data is sent to a central server where it is processed, interpreted, and broadcast to other Dash units, providing drivers with accurate information about traffic conditions. This mesh approach turns every Dash-equipped vehicle into a traffic sensor. Two-way Internet connectivity also opens the door for all sorts of new interesting features and services.

Subscription services

Many automotive GPS receivers support subscription services that provide real-time traffic, weather, and other driving information. These services work by sending special data over radio waves from satellites or ground-based stations. GPS units that use these information services have electronics that receive and display the radio data; the electronics can be built into a unit or sold as an add-on product.

A subscription service works like this. You pay a yearly subscription fee (generally, from $50 to $70, although some manufacturers are starting to offer access to traffic services free on their higher-end models). MSN Direct, XM NavTraffic, and Clear Channel TMC are the primary services in the United States, by the way.

After you enter an activation code in your compatible GPS unit, it starts receiving the data. An icon displays if there's traffic information for the road you're driving on or your current route. (There are icons for accidents, traffic congestion, construction, and other traffic annoyances.) You can get more information by selecting the icon and then decide whether to have your GPS unit suggest an alternative route around the problem.

Traffic information tends to be available only for major metropolitan areas. Be sure to check the service coverage area before you sign up to ensure you'll receive information for where you typically drive.

Although traffic subscription services sound pretty slick, user opinions have been, pardon the pun, all over the map. Some people swear by the services, others think they're of limited use; complaining of inaccurate information such as erroneous accident reports. This technology is still fairly new and, in my opinion, needs some time to fully mature. The good news is most manufacturers offer free trial subscriptions, so you can evaluate a service before you sign up for a year. All services aren't the same, and most GPS units support only one type of service. Considering this, I recommend doing some Internet research, checking service coverage areas, and reading about user experiences with a particular service you're considering.

Voice recognition

Some higher-end GPS units have a voice command recognition feature. You tell the receiver what you want to see or have it do. This eliminates touching the screen and distracting you from watching the road and concentrating on your driving.

Types of Automotive GPS Receivers

Now that you have an idea of some of the features of automotive GPS receivers, here's a look at the different types of products that are on the market. Automotive GPS units can fall into one of two categories, portable navigators and in-dash navigators.

Portable navigators

I call any automotive GPS receiver you can easily move between vehicles a portable navigator (one is shown in Figure 6-3). There are two types.

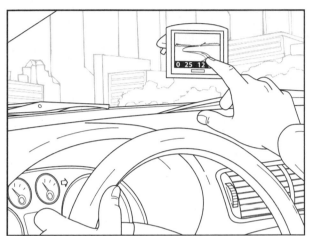

Figure 6-3:
Portable automotive GPS unit.

Full-size navigators

These units are larger, heavier. and have bigger screens. Some models don't have rechargeable batteries and only work when plugged into a car cigarette lighter.

If your GPS unit exclusively gets its power from a cigarette lighter cord, consider purchasing an adapter that converts 110-volt household alternating current (AC) to 12-volt direct current (DC). These handy converters plug into a wall electrical socket and include a cigarette lighter receptacle for plugging in car-powered devices. You can then use the GPS receiver in your house for entering addresses and planning trips.

Pocket navigators

Pocket navigators are compact GPS receivers that easily fit in a pocket or a purse. They have rechargeable Lithium-ion batteries, so you can use these units without having them plugged into a car cigarette lighter or other power source. That gives you the option of using them outside your car. In fact, some models will even calculate walking and bicycle routes for you.

Just because you can walk around with a pocket navigator doesn't mean it's equivalent to a typical handheld GPS receiver you'd use while hiking. Automotive GPS receivers only offer turn-by-turn, street navigation guidance and don't allow you to navigate in a straight line between Point A and Point B.

In addition, they don't support latitude and longitude waypoints, topographic maps, and other features you'd normally find in an outdoors GPS unit.

To address these limitations, some manufacturers offer "cross-over" automotive GPS receivers. These are hybrid pocket navigators that help you get around town but have some of the same features as handheld GPS units.

The biggest limitation to pocket navigators is their relatively short battery life when unplugged. Depending on the model and factors such as screen backlighting, expect a portable navigator battery to last anywhere from two to eight hours.

In-dash navigators

In-dash navigators are automotive GPS receivers built into a car or truck (an example is shown in Figure 6-4). Unlike their portable cousins, you can't easily move them from car to car. Often, these units are integrated with stereo systems.

There are two types of in-dash navigators:

✔ **Factory/dealer installed:** These GPS units come as options on new cars. Until several years ago, GPS systems were available only on luxury cars, but now they're becoming common in many makes and models (although you still pay a premium price for a factory unit).

✔ **Aftermarket:** With the popularity of GPS, most car stereo manufacturers are now offering models that combine navigation and entertainment systems. Here are Web sites for some of the top-selling manufacturers:

   • **Alpine:** www.alpine-usa.com

   • **Blaupunkt:** www.blaupunkt.com

   • **Kenwood:** www.kenwoodusa.com

   • **Panasonic:** www.panasonic.com

   • **Pioneer:** www.pioneerelectronics.com

I'm personally more of a fan of portable navigators because of their lower costs and portability. However, if you're interested in an in-dash GPS unit, your best bet is to visit a large chain electronics store or an automotive stereo retailer to learn more about them and get a demonstration.

Figure 6-4:
An in-dash
automotive
GPS unit.

Selecting an Automotive GPS Receiver

Before you purchase a GPS receiver, especially an automotive variety, you should kick the proverbial tires. Spend some time online comparing the many different brands and models to determine which one is going to best meet your needs.

I list a number of GPS receiver forums, blogs, and review Web sites in Chapter 20. I highly recommend reading through these information sources and checking out the skinny on any GPS model you're thinking about buying.

The top-selling manufacturers of automotive GPS receivers in the United States are Garmin, Magellan (a division of Thales Navigation), TomTom, Navigon, and Mio. All these manufacturers have extensive Web sites that provide detailed information about their products. If you're in the market for a GPS receiver, be sure to spend some time browsing through product literature. The Web site addresses for these manufacturers are

- ✔ **Garmin:** www.garmin.com
- ✔ **Magellan:** www.magellangps.com
- ✔ **TomTom:** www.tomtom.com

✔ **Navigon:** www.navigon.com

✔ **Mio:** www.mio-tech.com

The wise consumer just doesn't look at the marketing literature. Download the user manuals for the models you're interested in to better understand their features.

In addition to the top-selling brands, others brands of automotive GPS receivers are also on the market. In most cases, sticking with a known, name brand ensures a higher level of customer support.

After you've done your online homework, head out to a retail outlet that sells automotive GPS units. Most have displays with different models running demo software that simulates the GPS unit in use. This gives you a chance to get some hands-on experience seeing how the receiver works and better understanding its capabilities and features.

Automotive GPS receivers are like Windows, Mac, and Linux PCs. They all can do the same tasks, more or less, but have different user interfaces for getting the job done. Each GPS manufacturer takes a little different approach when it comes to entering commands and viewing information. I suggest you evaluate units from several different manufacturers to find the user interface that feels most comfortable to you.

I recommend three simple questions you should ask yourself before you begin your automotive GPS receiver search:

✔ **How much do I want to spend?** With today's prices in mind, you can spend anywhere from under $150 to over $1,000 on an automotive GPS receiver. As with handheld GPS units, the rule of thumb is the higher the cost, the more features you get. Keep in mind that if you don't need a lot of whistles and bells (how often will you be listening to music on your GPS unit?), a basic model may provide all the features you need and do just as good of a job navigating than a more expensive unit.

For the most part, the cost of a GPS receiver really has nothing to do with accuracy. More expensive units may have advanced, sensitive GPS chipsets though, which means satellite acquisition time is shorter and the receiver has an easier time locking onto signals.

✔ **How often will I use it?** Be honest with yourself, how often do you think you'll be using the GPS unit? If it's not very much, a basic model without a lot of features may suit you fine. Comparatively, if you're on the road a lot, a model that supports Bluetooth and provides traffic information just may make life better.

✔ **How will I use it?** Will you be using the GPS receiver in several vehicles? If so, pick a model that you can easily swap between cars. If you travel and frequently use a rental car, select a small, easy to pack model. If you want the flexibility of using the unit outside your car, be sure it has a rechargeable battery (and is a crossover model if you plan to use it like an outdoor handheld GPS receiver).

Using Your Automotive GPS Receiver

Congratulations! After wading through all of the available models and making an informed decision, you bought yourself an automotive GPS unit (or maybe you were lucky, skipped all of that, and received one as a gift). But what's next? In this section, I give you some practical advice on using your new GPS unit.

First things first

Open the box, admire your new toy, and read the user manual. You'll probably learn a thing or two.

If you can power your GPS unit from a wall socket or with a USB cable connected to your computer, turn it on and get familiar with it. (It's much more comfortable to do so in an easy chair compared to sitting in a car in your garage). Try using your GPS receiver near windows so it can have a better chance of locking onto satellite signals and mapping your current location so you can try out various features.

Give your GPS receiver's battery a full charge before heading out for a drive; if you can wait, that is.

Mounting options

Your next step is to mount your GPS unit in your car. This sounds simple, but there are a few considerations.

Your GPS receiver needs a clear view of the sky to work best. Most receivers have a built-in antenna to bring in satellite signals, although a few models support attaching an external antenna if the receiver body is somehow blocked.

This means you need to mount your GPS unit somewhere on or above the dashboard. (Even if your GPS receiver gets good satellite reception below the dashboard, I don't recommend you mounting it there. This creates a potential safety issue of looking down and taking your eyes off the road.)

Don't mount the GPS unit on the part of the dashboard where airbags deploy. In an accident, you don't want your GPS receiver turned into an unguided missile hurtling toward you.

Along with making sure your GPS receiver has a good view of the sky, you also need to ensure you have a good view of your GPS unit and that it doesn't obstruct your view of the road.

When you've found the best location for your GPS receiver, how do you mount it? You have several options.

Suction cup mount

The most common automotive GPS receiver mounting method is with a suction cup mount. This consists of a suction cup attached to a plastic cradle that holds the GPS unit (the cigarette lighter power cord plugs into the cradle on some models and directly into the receiver on others). A small lever firmly adheres the suction cup to the windshield; you can rotate and tilt the GPS unit on its mount so it's in an optimal viewing position. Press the lever in the opposite direction to remove the suction cup mount.

Here are a few mounting tips:

- Make sure the window glass is clean and dry before affixing the suction cup.
- Attach the suction cup to the straightest part of the window (the more curved, the less it will adhere) and press firmly.
- If there is a clear, thin piece of plastic on the suction cup, take it off before mounting. This is a protective piece of packaging.

Some states, such as California and Minnesota, have laws that make it illegal to have anything mounted to the front or side windows of a vehicle while it's being driven (other states are considering following suit). That means suction cup mount GPS units are a no-no, so be sure to check your state laws. You may need to use a dashboard mount to avoid a ticket.

Velcro

Another mounting option is to use Velcro, attaching it to the dashboard and the base of the GPS unit. Velcro has several disadvantages:

- You don't have much control over the screen's viewing angle, compared to a suction cup mount.
- You don't have the flexibility of moving the GPS unit to a different position unless you pull up the Velcro.
- Prolonged exposure to dashboard heat can weaken the Velcro adhesive.

Velcro has its place, but personally, I'd opt for another mounting option first.

Mat mounts

Some GPS receiver manufacturers offer an optional dashboard mounting system that uses a weighted mat with a plastic base. The inside of the base is smooth and provides a good surface for attaching a suction cup mount. The bottom of the base, which sits on the dashboard, has a high friction coating that prevents the mat from sliding around while a car is in motion. These are a great alternative to a windshield mount (especially in California and Minnesota).

Bracketron (www.bracketron.com) makes a mat mount called the Nav-Mat (shown in Figure 6-5) that works with most automotive GPS units. It retails for under $30 if you shop around.

Figure 6-5:
A Nav-Mat GPS receiver mat mount.

On the road

With your GPS unit now successfully mounted, let's go for a little drive.

I suggest keeping the user manual in the glove box for reference until you get the hang of using your GPS receiver.

Wait until your GPS receiver acquires satellites before starting your trip. A GPS unit determines its initial position faster when stationary versus moving. If you start driving and your GPS receiver hasn't locked onto satellites yet, the street map won't update while you drive. (A map will show your position right before the GPS unit was last turned off.) When signals from enough satellites are acquired, the map will refresh to your current location.

For your first outing with a GPS receiver, I suggest driving around just to get a feel for how it works. First, reset the trip log. When you get back home, check how far you went and your other trip log information.

It seems like every couple of weeks there's a news story about some hapless driver who followed directions from his GPS unit and ended up driving into a canal, going the wrong way on a one-way street, or crashing into a train. I can't emphasize this enough: You need to apply observation and common sense when using GPS for navigation. Don't believe your GPS receiver is infallible and follow it blindly.

After you're familiar with your GPS receiver, or if you're feeling adventurous on your first outing, try entering a destination address, having the unit calculate a route, and following its directions. It might suggest a different route than you normally take. Try it and see if it's smarter than you are.

Minimize the amount of time you play with your GPS receiver while driving. Ideally, you should set a route and not have to touch the GPS unit until you arrive at your destination. Primarily rely on the spoken directions and look at the screen only when required. Pull over if you need to fiddle with the receiver, or better yet, mount it in a location where your front seat passenger can operate it. That's what copilots are for, right?

After your initial road tests, go to the unit's preferences page and change any settings (display brightness, map detail, speaker volume, and so on) to make the receiver more usable. Don't be afraid of changing options. Many models have a restore default settings command that reverts all your changes to how the GPS unit came from the factory.

Updating maps and firmware

Your automotive GPS receiver is going to come with a built-in collection of street maps. The maps should be relatively current from whenever you purchased the GPS unit. However, because new roads are constantly being built as well as old roads sometimes changed, your maps will eventually become outdated.

Manufacturers sell updated map data that you can upload to your GPS unit; the new maps typically come on memory cards, CD-ROMs, or DVDs. Check your user manual or the manufacturer's Web site for information on how to upload new map data to your receiver.

Maps are often updated yearly with an upgraded version costing $100 and up (depending on the manufacturer and the map). More current POI data is often included with revised maps.

You can only use updated maps produced by your GPS manufacturer. For example, you can't load Magellan maps into a TomTom receiver.

Should you buy updated maps? Here's my advice:

✔ If you're using an automotive GPS unit for commercial purposes, such as deliveries or sales calls, it makes sense to purchase the yearly map updates — especially if you live in an urban area with lots of change and growth.

✔ If you're primarily using your GPS unit for recreational travel, staying on highways and visiting popular locations, there's likely not a big need for you to regularly purchase map updates. There probably haven't been enough changes to warrant buying revised map data.

In any case, be sure to save the receipt for your GPS unit. If new map data is available shortly after you purchase your GPS receiver, many larger manufacturers will provide the revised maps free with proof of purchase.

If you travel outside the United States for business or pleasure and are going to be driving, see whether your GPS manufacturer offers road maps for your destination that you can purchase and upload to your receiver.

Many GPS manufacturers, as well as the two primary street map data providers — Tele Atlas (`www.teleatlas.com`) and Navteq (`www.navteq. com`) — let users submit Web-based, map feedback reports. Reports are verified and eventually incorporated into map updates. If you encounter a map error, consider reporting it.

Updating your maps is optional, but I strongly recommend you update your GPS receiver's firmware when the manufacturer releases a new version. *Firmware* is the software that runs your GPS unit, and manufacturers often provide new versions, especially within the first two years of a product's life, that fix bugs and add new features.

Updating firmware is usually a straightforward process of downloading a file on your PC and then transferring it to your GPS unit with a USB cable or memory card.

Again, check your GPS manufacturer's Web site for information on how to determine which version of firmware your GPS unit is running, what the current version is, and instructions for installing the updated firmware.

Homework assignment

To wrap things up, here's a short homework assignment to see how well you know your automotive GPS receiver. It's an open book (user manual) test that you'll be grading on your own. Ready? Time to begin.

✔ Plot a route from your current location to 1600 Pennsylvania Avenue NW, Washington DC (20500). Write down how many miles away it is. What's the last turn you make according to the directions list?

✔ Write down the name of the nearest gas station POI and how far away it is.

✔ Change the measurement setting from statute (miles) to metric (kilometers) then back again.

✔ Change the screen display setting from daylight to night (or vice versa).

✔ Write down which version of the firmware your GPS unit is running.

✔ Enter this address in your address book. Wiley, 10475 Crosspoint Blvd., Indianapolis, IN 46256. (You can delete it when you're done, or save it if you ever want to stop by Dummies World Headquarters and say hello.)

Okay, pencils down. How'd you do?

GyPSies, tramps and thieves

During Christmas 2007, automotive GPS receivers became one of the season's hottest electronic gifts. And because of their popularity, portable car navigation systems continued to be hot (as in stolen) in the months that followed. Because the units are easy to steal and resell, automotive GPS receiver theft is becoming a big problem. It's so bad in some places that police are setting up sting operations to catch thieves. Here are a few tips to avoid becoming the victim of theft:

✔ **Lock your door.** This is a no-brainer (although it doesn't prevent a thief from smashing a window).

✔ **Secure or take your GPS unit with you.** Out of sight is out of mind. Consider that shopping mall, restaurant, and hotel parking lots are common crime scenes.

✔ **Clean your window.** According to police, thieves often look for a telltale suction cup mark on a car's window. Even if the GPS unit and mount aren't visible, the round smudge may cause a thief to break into the car anyway to look in the glove box or under the seat. Wipe off the mark or use a removable mat mount that sits on your dashboard.

✔ **Report the crime.** There's evidence to suggest GPS theft statistics may be even higher because people aren't bothering to report the crime.

✔ **Record your serial number.** Most people don't bother recording serial numbers. Although knowing your unit's serial number won't prevent theft, it's useful for insurance claims and may help get the GPS receiver back to you or help catch the person that stole it.

A few third-party cable locks for GPS units are on the market, but ultimately, if the theft problem continues to escalate, it will be up to the manufacturers to incorporate solutions. The car stereo industry had to respond to a similar problem, and started offering removable faceplates and technology that disables a radio when it's removed from a car. You're starting to see some GPS receivers with optional personal identification numbers that must be correctly entered before the unit will function. Expect new and innovative security features to be available in the near future.

Chapter 7

Cellphones, PDAs, and Other GPS Devices

In This Chapter

▶ Navigating with GPS cellphones

▶ Using PDAs with GPS

▶ Tracking people, pets, and vehicles with GPS

▶ Recording GPS information with data loggers

▶ Learning about GPS radios

*W*ithin the past several years, GPS seems to be showing up everywhere. Originally found in specialized devices used by boaters, pilots, and hikers, GPS technology is now commonly found in cars (which I cover at length in Chapter 6), cellphones, and a variety of other electronic gadgets; some you may not even known about.

In this chapter, I fill you in on GPS-enabled gadgets, including cellphones, Personal Digital Assistants (more commonly known as PDAs), tracking devices, data loggers, and other GPS devices such as cameras and radios.

Because most everyone has a cellphone these days, I start by telling you everything you need to know about how mobile phones work with GPS and serve as PNDs (that's Personal Navigation Device if you're not hip to the marketing acronyms and buzzwords).

GPS and Cellphones

Using a cellphone to find where you're at is a little like magic. You run a program, a map pops up, and there you are. You can even type in an address and receive turn-by-turn directions on how to get to your destination. Just like a handheld or automotive GPS unit.

Well, not exactly. Although many cellphones seem to offer the same capabilities that dedicated GPS units do, there are some important differences between the two devices.

In this section, I explain different technologies cellphones use for providing location information (you might be surprised to learn not all of them rely on GPS). I also talk about location-based services that are available for many cellphones and point out key differences between phones and their outdoors and automotive GPS navigator relatives.

Cellphones and Assisted GPS (A-GPS)

In 1996, the federal government mandated that cellular phone carriers had to pass on the location of a mobile phone that dialed 911 to emergency call centers; just like how a 911 call made from a landline phone provides a street address. Although GPS had long been operational, it wasn't deemed practical to use in cellphones because of many reasons — receiver cost, size, power consumption, length of time to acquire a satellite signal, and problems locking on to satellites in cities, canyons, and under trees.

To meet the Enhanced 911 requirement for wireless phones, T-Mobile and AT&T (formerly Cingular) came up with ways to determine a phone's location by triangulating its position based on cellular network towers.

Carriers such as Verizon Wireless and Sprint-Nextel met the requirement by building a special GPS chip into most of their phones. These chips only needed data from two GPS satellites (versus a minimum of three for a conventional GPS receiver). The wireless network itself, using "assistance servers" broadcast additional GPS information required to fix a location. This system is known as *Assisted GPS* or A-GPS.

Compared to GPS chips found in handheld and automotive receivers, A-GPS chips are cheaper, use less power, lock on to an initial satellite faster (called *Time To First Fix*), and can provide location data in places a conventional GPS receiver might have difficulties working, such as indoors or in urban areas with tall buildings. A-GPS chips are also fairly accurate, from within 5 to 50 meters.

The two most important things to know about A-GPS are

- ✔ Most A-GPS cellphones can only provide your location when you have wireless network coverage. (Some smartphones can run in standalone GPS mode without needing a cellular network.)

- ✔ The cellular carriers call the shots on what type of location services you can access with your A-GPS phone. Typically, there are additional fees and service plans. I talk about these in the upcoming "Location-based services" section.

The Qualcomm GPSOne, currently the most popular A-GPS chip, can be configured to operate in standalone GPS mode without assistance servers and works just like a handheld GPS receiver. However, it's up to a carrier to enable this mode. Depending on the phone model, it might be possible to do a little hacking to enable standalone mode. Some Googling can provide instructions, but you're on your own the minute you start using undocumented commands (something cell providers tend to frown on).

If you have a compatible Nokia phone, check out the free Sports Tracker application (http://sportstracker.nokia.com), which saves GPS track files and provides speed and distance information.

Bluetooth cellphones and GPS

Another option for using GPS with a cellphone, that doesn't rely on A-GPS chips or require monthly fees, is to pair a Bluetooth compatible phone with a Bluetooth GPS receiver.

In a nutshell, the Bluetooth receiver wirelessly transmits current GPS position data to the phone, while navigation software loaded on the phone plots your location on a moving map.

I cover Bluetooth GPS receivers and map programs in the upcoming "GPS and PDAs" section.

Non-GPS cellphone navigation

You don't necessarily need an A-GPS chip in your cellphone to use it as a navigator. For example, Google Maps Mobile (www.google.com/gmm) is a free service that works on most Internet-enabled phones. It provides a moving map showing your current position as well as directions to a street address. Instead of using GPS, the service relies on information received from cell tower sites to determine your location (Google Maps can also use GPS directly on some phone models). It's not as precise as GPS, with accuracy of roughly 1,000 meters, but you can't beat the price.

In addition to information from cell towers, the Navizon location program for older models of the popular Apple iPhone uses data from WiFi nodes to help zero in on a location — newer iPhones have an A-GPS chip. Even with the additional WiFi data, don't expect pinpoint accuracy but just a general idea of where you are.

Google Maps Mobile and similar services retrieve maps from the Internet so you'll need some type of a data plan with your phone. Consider an unlimited plan so you aren't stuck with a big download bill.

Location-based services

Location-based services (LBS) is a trendy marketing term for any service that wirelessly provides certain types of information to a cellphone (or PDA or laptop), based on its location. LBS depend on GPS, GSM, GPRS, and a whole bowl of other alphabet soup acronyms. (Don't worry; I'm not going to get into them.)

Some examples of common location-based services are

- Getting turn-by-turn directions to a restaurant
- Requesting the location of the nearest ATM
- Displaying sale ads or coupons for a nearby retail store
- Seeing where friends or family are on a real-time map
- Receiving traffic or weather information

Two general types of location-based services are available:

- **Carrier provided:** Services offered and provided by the cellular carrier. Navigation service for compatible cellphones within a network is one example.

  The cost of navigation services depends on the carrier, but figure around $10 per month or less. If you don't regularly need your cellphone to act as an electronic map, but are heading to a city you've never visited before, cellular providers often enable navigation services on your phone for a few dollars a day. Check your carrier's Web site for details.

- **Third party:** Services provided by third parties that typically require your cellphone to connect to the Internet to access data. You might need to load a program on your phone to use the service. In addition to a cellphone data plan, you may also need to pay a subscription fee to the service provider.

Some popular location-based services include

- **Earthcomber:** www.earthcomber.com
- **Garmin Mobile:** https://my.garmin.com/gmbg/start.faces
- **loopt:** http://loopt.com
- **MapQuest:** http://wireless.mapquest.com
- **Telenav:** www.telenav.com
- **uLocate WHERE:** www.ulocate.com

If you're investigating location-based services for your cellphone, the first thing to do is check whether your phone will work with the service. Service providers list compatible phones on their Web sites.

Cellphones versus dedicated GPS units

Just so we're on the same page, the primary purpose of a cellphone is to, drum roll please, serve as a portable telephone. All the other whiz-bang features, including navigation, are secondary purposes, designed for customer convenience (and to sell more phones and services). In most cases, compromises must be made in shoehorning secondary features into a small cellphone.

So the big question is, if your cellphone has GPS functionality, can it totally replace a dedicated GPS receiver? Not really, and here's why:

✔ Size and cost prevent a complete set of handheld or automotive GPS features from being added to a cellphone.

✔ Unless the phone has standalone GPS capabilities, it will not be as accurate as a handheld or automotive GPS unit (that means no geocaching with your phone).

✔ For outdoor use, handheld GPS units are waterproof and considerably less fragile than cellphones.

✔ Most A-GPS phones depend on cell service to provide location information. No cell service usually means you don't get your current position.

✔ Cellphone screens are pretty small for road navigation and lack large, easy-to-use touch screen controls found in portable automotive GPS receivers.

✔ Cellphone location services typically have monthly fees compared to dedicated GPS receivers, which provide location information for no cost after you purchase them.

Cellphones with location-based services are convenient because most people usually have a cellphone with them. They're especially useful in urban locations when you're on foot, looking for an address. However, I don't see cellphones as a complete replacement for handheld GPS units in the outdoors or for automotive GPS systems on the road.

If you want to keep current on cellphone/GPS technology, including tracking the latest models, check out the Cell Phone GPS blog at `http://cellphonegps.blogspot.com`.

GPS and PDAs

Cellphones that do just about everything have cut into the PDA market in recent years, but new Palm and Pocket PC devices are still being sold and can make pretty good portable navigation devices when paired with GPS. In this section, I provide information and advice on using GPS with PDAs.

Choosing between a handheld GPS receiver and a PDA

You might wonder why anyone would want to use a PDA instead of a hand-held GPS receiver. That's a very good question; you gadget junkies out there who have your hands raised and are answering, "Because it's cool," please put your hands down and continue reading.

You'll find compelling advantages and disadvantages to using a PDA with GPS that are based on your intended use and needs. To see whether you should even be considering a PDA navigation system, review some of the pros and cons right from the start.

PDA advantages

Aside from being cool, a PDA might make sense for you as part of a personal navigation system for a number of reasons. Some of the advantages include

- **Larger screens:** PDAs have larger, higher-resolution, color screens compared with most handheld GPS receivers. This is a big plus if your eyesight isn't as good as it used to be — and it's really important if you're using the PDA while driving. You want to be able to quickly glance at a map on the screen, determine your location, and then get your eyes back on the road.

- **More maps:** Most of the maps that you can upload to GPS receivers don't have a lot of detail, especially the topographic maps. These maps tend to be vector line maps and don't have the resolution or detail found on paper maps that you'd use for hiking. Several mapping programs are available for PDAs that support all types of maps, and you can even create your own custom maps. With a PDA, you can use more detailed maps, such as scanned, color 1:24,000 topographic maps. **Bonus:** You're not locked in to using only a GPS receiver manufacturer's proprietary software and maps.

- **Expandable memory:** Unlike older GPS receivers, which have fixed amounts of memory, most PDAs support expandable memory with plug-in memory cards. The only limitation to the number of the maps and amount of data that you can store is the size of the memory card. (GPS receivers that use memory cards share this advantage.)

✓ **Usability:** Although handheld GPS receivers are fairly easy to use, the user interfaces found on PDAs are even simpler. Using a touch screen to enter data and commands is a lot faster and easier than using the buttons on a handheld GPS receiver.

✓ **Inexpensive:** Used PDAs that make great GPS platforms are relatively cheap on eBay and other online sources.

✓ **Custom programs:** Developers can easily write custom programs for PDAs that access the data output from a GPS receiver. If you're collecting information that's based on location data, this can make your job much easier than pressing buttons on a GPS receiver and then handwriting remarks in a field notebook.

✓ **PDA features:** PDAs have all sorts of useful programs such as address books, contact lists, and databases designed for readily storing data. A fair amount of this information tends to be location based (like addresses), and having a single information/navigation device is the definition of practical.

PDA disadvantages

After reading through advantages of using a PDA as your navigation system of choice, you're probably sold on using a PDA. However, they definitely aren't for everyone. Some downsides include

✓ **Ruggedness:** Handheld GPS receivers are designed to take more abuse than PDAs, which often fail when they're dropped or knocked around. Although you can buy *ruggedized* (with special enclosures that make them waterproof, drop-proof, bear-proof, and kid-proof) PDAs, they're considerably more expensive than off-the-shelf models; expect to spend at least several hundred dollars more.

✓ **Weather/water resistance:** Unlike GPS receivers, PDAs aren't designed to be waterproof or even weatherproof. This can be a major issue if you plan to use your PDA navigation system outdoors in damp, rainy, or snowy weather, you're around water, or you have a leaky water bottle in your backpack.

If you use your PDA in the great outdoors, I recommend a waterproof, crush-proof carrying case. I like OtterBox products; you can learn more about them at www.otterbox.com.

✓ **Power considerations:** Most PDAs use internal batteries that recharge through a docking cradle. If you're away from a power source, this can be a serious issue because you can't swap out dead or dying batteries for a convenient set of spare AA or AAA batteries like you can with a handheld GPS receiver.

When it comes to weighing the pros and cons of PDA navigation systems, you really have to examine your needs and planned use. If you plan to use a GPS receiver exclusively for road navigation, you might consider a PDA as a multi-purpose alternative to a car navigation system. However if you're going to be using GPS primarily in an outdoor setting, you're probably better off with a handheld GPS receiver.

Interfacing your PDA to a GPS receiver

If you've decided a PDA navigation system meets your needs (or maybe you just love cool high-tech toys), the first step is to decide how you're going to get GPS data into your PDA. Options for doing so include

- ✔ Handheld GPS receivers connected to the PDA with a serial (or USB) cable.
- ✔ *Mouse* GPS receivers (a GPS receiver with no display screen and a serial or USB cable).
- ✔ GPS receivers built into PC or memory cards.
- ✔ Wireless GPS receivers that transmit data with Bluetooth radio signals.

Mouse, card, and Bluetooth GPS products may have older GPS chips or newer technology, high-sensitivity chips. Opt for a high-sensitivity model if you want faster satellite locks and better reception in urban and wooded areas.

I'm assuming you already have a PDA or are currently shopping for one and have a pretty good idea what you're going to buy. It's outside the scope of this book to make suggestions on which PDA you should use. (Plus, I'm not going to pick sides in the Palm versus Pocket PC/Windows Mobile brand-loyalty wars.) Just keep in mind that most PDAs can interface with a GPS receiver one way or another.

Most of the GPS receivers designed for use with PDAs cost as much as low-end to mid-range handheld GPS receivers. Some also work with laptop computers, which provide the ultimate big-screen GPS display.

Handheld GPS receivers

If your PDA has a serial port, and your GPS receiver uses a serial cable to connect to a PC, you're in business. Most of the information on connecting a personal computer to a GPS receiver that I discuss in Chapter 10 also applies to PDAs.

Because you're using the handheld GPS receiver exclusively as a data input source — and really don't care about what's appearing on its screen — just about any GPS receiver that can communicate with a computer via a serial port will fit the bill (older, used GPS units can be found pretty cheaply these days). You don't need a lot of features and whistles and bells on the GPS receiver if you're primarily using it this way. This option makes a lot of sense because the GPS receiver can be used independently of the PDA, especially outdoors during bad weather.

The primary disadvantage is that you have to contend with two electronic devices and the cable that connects them together. This can sometimes get a bit messy in a car, with hardware and power and interface cables scattered all over your dashboard. Also, this type of PDA navigation setup is a bit cumbersome to deal with if you're walking around.

One of the best sources of GPS receiver interface cables for a wide variety of PDAs is Pc-Mobile. Check out its extensive product Web site at http://pc-mobile.net.

Mouse GPS receivers

GPS receivers that don't have a display screen but connect to a computer with a serial or USB cable are often called *mouse* receivers because of their resemblance to a computer mouse. A mouse GPS receiver acts as an input device for a PDA or laptop and only sends satellite data that it's currently receiving.

Mouse GPS receivers are about half the size of the smallest handheld GPS receivers (Figure 7-1 shows the size differences between several GPS receivers), but even so still have good satellite reception with open skies. Another advantage to a mouse GPS receiver is you can place it in an optimal spot on your car's dashboard for satellite reception and then mount the PDA in the most visible place for driving. Depending on the model and type, a mouse GPS receiver can be powered by a cigarette lighter adapter, rechargeable batteries, or the device it's plugged in to.

Although mouse GPS receivers are smaller than handheld GPS receivers, you still face the cable clutter issue; plus, you can't use the mouse receiver to get satellite data unless it's connected to a PDA or laptop.

Some of the vendors of mouse GPS receivers (and their Web sites where you can get product information) include

- **DeLorme:** www.delorme.com
- **Deluo:** www.deluogps.com
- **Garmin:** www.garmin.com
- **Haicom:** www.haicom.com.tw
- **Holux:** www.holux.com
- **Pharos:** www.pharosgps.com

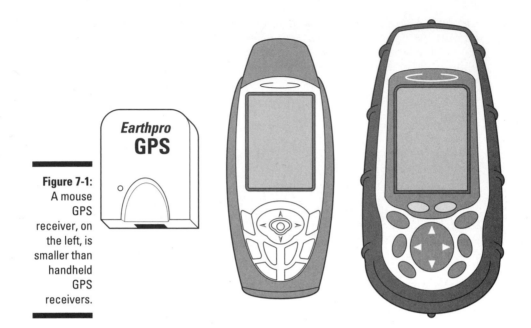

GPS receiver cards

Some GPS receivers take the form of a card that you can plug into a PDA expansion slot. The two types of GPS receiver cards are

- ✔ **Memory cards:** Most PDAs have a memory card slot that supports a Compact Flash (CF) or Secure Digital (SD) type of memory card. Both of these card formats also support hardware devices that can be embedded into the card: in this case, a GPS receiver. (PDAs without internal memory card slots might have optional *expansion packs* that plug into the PDA and provide a memory card slot.)

- ✔ **PC Cards:** A PC Card (also called a PCMCIA card) is a hardware expansion card designed for laptops. PC Card devices include hard drives, modems, and GPS receivers. These receiver cards are mostly used with laptops, but some PDAs support expansion packs for using PC Card devices.

Using a GPS receiver card with your PDA means that you don't need to worry about an external GPS receiver or cables. You just plug the card into a slot, and the GPS receiver starts sending satellite data to a navigation program. However, note these drawbacks:

- ✔ **Slot competition:** CF and SD memory cards that function as GPS receivers take up the expansion slot that's used for additional PDA memory.

 ✔ **Battery hogs:** GPS receiver cards quickly run down your PDA's battery if you're not connected to an external power source.

Some of the main GPS receiver card manufacturers (and their Web sites) are

 ✔ **Deluo:** www.deluo.com

 ✔ **GlobalSat:** www.globalsat.com.tw/eng/index.htm

 ✔ **Haicom:** www.haicom.com.tw

 ✔ **Holux:** www.holux.com

 ✔ **Pharos:** www.pharosgps.com

GPS cards also work great with laptops. Additionally, USB plug-in GPS receivers are available.

Bluetooth GPS receivers

Of all the ways to connect a GPS to a PDA (or laptop or cellphone for that matter), my absolute favorite is to use a Bluetooth GPS receiver. Because most PDAs and laptops support Bluetooth these days, it's just a matter of installing some software, configuring your PDA, and turning the Bluetooth GPS unit on. Just like magic, GPS data is wirelessly transmitted. Bluetooth GPS receivers are pretty slick because they don't

 ✔ **Rely on the PDA's batteries for power**

   They have their own power source and will run from 6–10 hours when fully charged, depending on the model.

 ✔ **Use one of the PDA's expansion slots**

 ✔ **Need cables that can become tangled**

   Bluetooth devices have about a 30-foot range, and the GPS receiver can be placed in an optimal position on the dashboard to receive satellite signals.

Just place your Bluetooth GPS receiver (about the size of a mouse GPS receiver) anywhere with an open view of the sky, and it will broadcast GPS data to your Bluetooth-enabled PDA. (If you have an older PDA that doesn't support Bluetooth, there are Bluetooth receivers that plug into your PDA's memory card slot.)

Some Bluetooth models serve dual duty and allow you to connect the GPS receiver with a USB cable.

Many Bluetooth GPS receivers are on the market (most of the manufacturers previously listed offer one or more models). Your best bet to learn more is to do a Google search for *bluetooth gps.*

For outdoor use, you can easily mount your Bluetooth GPS receiver on a high spot, such as on top of a pack (or some other location that's in an optimal position to receive satellite signals) and wirelessly record GPS data with your PDA. OtterBox (www.otterbox.com) also makes waterproof cases for Bluetooth GPS receivers that don't degrade the transmitted radio signals so you can create a rugged wireless PDA navigation system for use in harsh conditions.

The integrated-GPS-receiver-in-a-PDA market never really took off, so if you're looking for a PDA navigation solution, some great deals on discontinued Garmin iQue models can be found on eBay and elsewhere.

For an extensive information on GPS receiver devices available for Pocket PC PDAs, including reviews and detailed specifications, check out www.gpspassion.com.

Reviewing PDA mapping software

If you buy a GPS receiver specifically designed for PDA use, it may come bundled with mapping software that includes map data and a program that displays the maps and interfaces with the GPS receiver.

Many companies that offer PDA mapping programs also sell versions that run on cellphones.

In addition to the program you install on your PDA, a program that runs on your PC loads the maps onto your PDA and can be used for route planning. You select the maps that you want to install on your PC and then upload the selected map data to your PDA. Because the maps are typically stored on the PDA's memory card, the larger the card, the more maps you'll be able to use.

Most PDA navigation software is designed for street navigation (see Figure 7-2 for screenshot of a typical PDA street map program) and has features for getting around on roads and highways, including

- **Autorouting:** By inputting starting- and ending-point addresses, the map program creates a route for you to follow to reach your destination. (You usually can choose between fastest or shortest routes.) The route is outlined on the map, and the program provides turn-by-turn directions to get to your destination.

- **POI data:** In addition to maps, most programs have extensive databases of POIs (Points of Interest) information, including gas stations, restaurants, shopping locations, and other useful travel data. POIs appear as icons on the map that you can click to get more information. You can also search for specific POIs by geographic location.

✓ **Real-time tracking:** When your PDA is connected to a GPS receiver, an arrow moves on the screen, giving you real-time information about your current position and speed as well as displaying where you've been.

✓ **Voice prompts:** In addition to displaying turn-by-turn directions on the screen, many programs provide voice prompts that tell you when to make turns to reach your final destination. This is a nice safety feature because you can pay more attention to the road and less attention to the PDA screen.

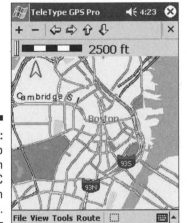

Figure 7-2:
Map display from Pocket PC navigation software.

A number of PDA mapping programs are on the market, and I could easily devote an entire book to discussing them all. Because this chapter provides only a general review of using PDAs with GPS, Table 7-1 lists of some of the more popular PDA mapping programs, including which types of PDAs they work with and a Web site address to get more product information. These programs are sold individually or come bundled with a GPS receiver. Expect to pay anywhere between $40–$150 for a software package (more with an included GPS receiver).

Most of the software and maps that come bundled with PDA GPS receivers are designed for road navigation. But what if you want to display topographic or nautical charts on your PDA or use real-time GPS tracking off the road?

You're in luck because several PDA programs fill this need. Topographic map display is one area that PDA mapping software can exceed handheld mapping GPS receivers. Your PDA can display full-color, detailed 1:24,000 scale maps that look exactly like the USGS paper versions. Compared with the 1:100,000 scale vector maps typically used on mapping GPS receivers, there's really no comparison when it comes to the amount of map detail that a PDA can display.

Table 7-1	Popular PDA Mapping Programs		
Program	**Pocket PC**	**Palm**	**Web Site**
DeLorme Street Atlas USA Handheld	X	X	www.delorme.com
Destinator	X		www.destinator technologies.net
Intellinav	X		www.intellinav.com
Garmin Mobile 10	X	X	www.garmin.com
Microsoft Pocket Streets	X		www.microsoft.com
IGuidance	X		www.inavcorp.com
OnCourse Navigator	X		www.oncourse navigator.com
TomTom Navigator	X		www.tomtom.com

If your journeys take you off the beaten path, here are some products you should be aware of:

✔ **FUGAWI:** FUGAWI was one of the first Windows desktop mapping programs and now works with Pocket PCs and Palms. The product comes bundled with U.S. street maps and nautical charts, or you can import maps of your own, such as USGS Digital Raster Graphics topographic maps. For additional product details, go to www.fugawi.com.

✔ **GPS Tuner:** A Pocket PC product (requires Windows Mobile 5 or 6) that supports Google Earth, TerraServer, and custom JPG maps. For more information, visit www.gpstuner.com.

✔ **Maptech Pocket Navigator:** The Maptech Pocket Navigator is a Pocket PC product compatible with USGS topographic maps and NOAA nautical charts. You can find more about it at www.maptech.com.

✔ **OziExplorerCE:** OziExplorerCE is a Pocket PC version of the popular OziExplorer mapping program. The CE version allows you to import maps that you've created with OziExplorer on your PC to your PDA. See more about the PC version OziExplorer in Chapter 16. For additional information on the CE edition, go to www.oziexplorer.com.

✔ **Pocket TOPO!:** An add-on product for National Geographic's popular TOPO! series that loads separately purchased TOPO! maps onto your Palm PDA. For more information, do a Google search for "Pocket Topo."

✔ **TeleType GPS:** Although originally designed for street navigation, TeleType's PDA mapping program supports importing aerial photographs and digital maps from the TerraServer-USA Web site. For more information on the software, visit `www.teletype.com`.

Here are several excellent Internet resources that can provide you with additional information on PDAs and GPS:

✔ **Dale DePriest's Navigation and GPS Articles:** Dale DePriest is a long-time contributor to the `sci.geo.satellite-nav` USENET newsgroup and the comprehensive `gpsinformation.net` Web site. He has a considerable amount of practical information on using Palms, Pocket PCs, and GPS at `www.gpsinformation.org/dale`.

✔ **Pocket GPS World:** Check out this European Web site for news, information, and forums devoted to mobile GPS products. The site has an extensive review section of PDA GPS receivers and is located at `www.pocketgpsworld.com`.

✔ **GpsPasSion:** This European site covers mobile GPS devices (with a lot of PDA information) and includes forums, links, news, and more. To visit the site, go to `www.gpspassion.com`.

GPS Trackers

Remember those old spy movies where the hero would place a bug under the bumper of a bad guy's car and then follow him using a hokey looking, small television set? Now anyone can be James Bond, thanks to GPS and cellular phone networks (and you don't even need the cheesy TV set).

GPS trackers allow you to track people, cars, boats, and even pets. The devices rely on a GPS chip connected to a transmitter that sends location data (often over a GSM/GPRS cellular phone network) to a tracking company's server. You log onto the company's Web site, and presto, whoever or whatever you're tracking appears on a moving map.

Any GPS tracking product that sends coordinates over a cellular network has monthly fees associated with it. Before you purchase a tracker, be sure to read the fine print.

In this section, I briefly discuss different types of GPS trackers.

People trackers

A growing number of GPS products designed to track people are on the market. Their uses range from law enforcement officers tracking bad guys to a parent wanting to always know where a small child is to friends simply keeping track of where other friends are.

People trackers fall into two general categories.

Dedicated trackers

These are small electronic devices with the sole purpose of receiving GPS data and then transmitting the coordinates in real time to a monitoring service. Many of these products are designed for covert use and have features such as magnetic bases for mounting on a car, A-GPS chips to allow indoor tracking, and extremely long battery lives. (A small, covert GPS tracker is shown in Figure 7-3.) Tracker prices can range from $400 to several thousand dollars.

BrickHouse Security is currently the largest supplier of GPS tracking devices. You can check out their products at www.brickhousesecurity.com.

It's estimated that over 700 police departments and agencies in the United States are using GPS tracking devices for investigative purposes. In 2007, the 7th Circuit Court of Appeals ruled police don't need a search warrant to use a GPS tracker. Welcome to 1984!

Cellphones

Although you may not have thought about it, the technology inside a cellphone makes it an ideal tracking device. In fact, most of the dedicated trackers mentioned above are simply miniaturized, GPS-enabled cellphones without a microphone, speaker, keypad, and display.

Because a cellphone can report its position with or without GPS, cellular carriers saw a business opportunity, and along with third parties, now offer services that will track the location of a cellphone. A number of these services target parents who want to keep track of their teenagers; a kind of Big Mother is Watching thing.

Figure 7-3:
A covert
GPS tracker.

Here's a sampling of some cellphone-based tracking products:

- ✔ **Accutracking** (www.accutracking.com): Provides tracking services, including speed and location, for compatible phones. Under $10 per month (not including data plan costs).

- ✔ **MapQuest Find Me** (www.mapquestfindme.com): MapQuest's Find Me service allows you to view the location of Find Me subscribing phones on a map or track a subscriber's movements. The service costs $3.99 per month, not including data plan, on select Sprint/Nextel and Boost phones.

- ✔ **Wherify** (www.wherify.com): Wherify offers the Wherifone, an easy-to-use GPS locator phone designed for children and senior citizens. The phone is priced under $100, with monthly plans starting at $20.

Check your cellular provider's Web site to see what types of people locator services they offer.

Pet trackers

For many people, a dog or a cat is as much a part of the family as children. And if Fido or Bucky the Cat goes missing, that's just too much to take. Entrepreneurs have responded to this need by offering GPS pet tracking products. These devices attach to a pet's collar and send location data either over a cellphone network or via radio signals to a handheld receiver. A pet tracker is shown in Figure 7-4.

Figure 7-4:
A GPS pet
tracker.

Most pet tracking systems allow you to define a range for your pet, sort of like an electronic, invisible fence. If Fido leaves the area, you're alerted via e-mail or text message.

Here are a few pet trackers with links to manufacturer Web sites:

  ✔ **Garmin Astro Dog** (www.garmin.com): A dog tracking system that sends GPS coordinates with VHF radio signals, meaning no monthly fees or cell service is required. Designed primarily for hunting dogs with a range of up to 5 miles; the dog's location displays on a full-featured, handheld GPS receiver.

  ✔ **Global Pet Finder** (www.globalpetfinder.com): A GPS/cellular network tracker designed for dogs over 30 pounds. A monthly service fee is required.

  ✔ **RoamEO Pet Tracker** (www.roameoforpets.com): Transmits a dog's position using radio signals, with range limited to about a mile. The location appears on a handheld color monitor. No monthly fee is required.

See SPOT run

One of the cooler GPS gadgets to appear on the market is a small orange box called SPOT Messenger. Press a button and SPOT uses a Globalstar satellite phone transmitter to transmit your coordinates from just about anywhere on earth.

Similar emergency distress devices for boats, called EPIRBs (Emergency Position Indicating Radio Beacons), and for people, called PLBs (Personal Locator Beacons), send a signal to government search and rescue organizations.

SPOT does the same, but also features a non-emergency button you press just to tell people where you are. The coordinates are transmitted to a corporate Web server and then an e-mail is sent to people you've pre-selected. The e-mail contains a link that displays a map with your current location. You can't send a text message with the coordinates, but it's a nifty way to check in with Mom on your way to the top of Mount Everest.

SPOT is still fairly new technology. With any new life-safety device, I always recommend a wait-and-see period before putting your full trust in it. A good place to check what the pros think of SPOT in ongoing evaluations is Doug Ritter's excellent Equipped to Survive Web site (`www.equipped.org`).

The SPOT Messenger retails for under $150, and a year's subscription to the service is around $100. For more information, visit `www.findmespot.com`.

✔ **Stealth Pet Tracker** (`www.surveillance-equip.com/pet.gps.html`): The size of most pet trackers makes them more practical for dogs — the bigger the puppy, the better. The Stealth Pet Tracker is one of the smallest products on the market, built for cats and smaller dogs. (Other manufacturers are working on smaller versions of their full-sized trackers, so expect to see more choices.)

GPS pet trackers cost in the $200 to $600 range (plus any additional monthly service fees), and like most GPS products, the more features, the higher the price. Do a Google search for *GPS pet tracker* to get additional information and see other products.

Vehicle and vessel trackers

I want to mention that GPS is widely used by trucking and freight companies and other businesses that have a large number of vehicles on the road. Vehicle tracking allows businesses to know where their vehicles are for dispatch purposes, view real-time speeds, and cut costs by selecting the most efficient routes. Because vehicles may be out of cellular phone coverage areas, commercial tracking systems often relay coordinates to satellites, which are then passed on to monitoring sites. A variety of fleet management products are available; I suggest you Google *fleet management gps* to get additional information.

If you want to know about using GPS to track a single vehicle, see the "People trackers" section earlier in this chapter.

Of more interest to the average reader, especially if you're a boater, is an up-and-coming, GPS-related tracking technology called AIS, which stands for Automatic Identification System. AIS is a little black box that broadcasts a radio signal with identification data and GPS coordinates (via either a built-in GPS receiver or a connected, external GPS device). It's very much like an aircraft transponder that towers and air traffic control centers use to track who's in the sky and where. The biggest difference is that with an AIS receiver hooked up to a laptop running compatible navigation software, anyone can see where other vessels using AIS are located (as well as get additional information about the vessel). Commercial ships are already using AIS, and the technology is the literal wave of the future for recreational boaters in coastal waters. For more information about AIS, go to `www.milltechmarine.com/resources_links.htm`.

For a cool demonstration of AIS in action, visit `www.shinemicro.com/Live.asp`, which shows live AIS-equipped vessel traffic from all over the world. (You need to register for a free account to view the charts.)

GPS Data Loggers

GPS data loggers are devices that collect *tracks* (the digital breadcrumbs of where a GPS receiver has been). Unlike a handheld GPS unit, a data logger isn't meant for navigation; although a few models have Bluetooth interfaces that send current coordinates to compatible devices. You can clip a logger to your belt or toss one in the top of a backpack, and the little gadget with no screen (one is shown in Figure 7-5) saves GPS coordinates to its memory. When you get home, you can download the tracks to a PC.

GPS data loggers serve two purposes:

- **Mapping:** Recording track points (coordinates, time, and elevation) that can later be imported into a map program.
- **Geotagging:** Associating GPS coordinates with digital photos; also sometimes called *geocoding* (see the "Geotagging in a Nutshell" sidebar for details).

After you've geotagged a photo, if you want to print the coordinates and time directly on the image, get a copy of Mike Lee's GPStamper (`http://losslessjpegtoolbox.wordpress.com`).

Figure 7-5:
A GPS data
logger.

GPS data loggers have a number of common features, including:

 Maximum number of track points: Most models save track points to internal memory and can store anywhere from 50,000 to over 500,000 points depending on the model (compare that to handheld GPS receivers that can typically only save up to 10,000 track points to internal memory).

✔ **Track point recording interval:** All models allow you to set how often a track point is saved, usually by time or by distance. For example, if you set the logger to save a track point every 15 seconds, some simple math calculations will tell you how many hours of data can be stored based on the device's maximum number of track points.

 When you're logging data in a car, you're speeding along and covering a fair amount of distance. For good mapping accuracy, be sure to set the device to record track points more often compared to when you're walking.

✔ **Power source:** Different data loggers use different power sources including rechargeable Lithium-ion batteries and standard AA and AAA batteries. Several models have small solar panels to provide supplemental power. Product specifications tell you how long the device will run before it needs more juice.

Geotagging in a nutshell

Digital camera photos saved in JPEG format have a series of tags that store information about when you took the photo, camera settings when you pressed the shutter, and additional description data. This is EXIF, or Exchangeable Image File Format, data. EXIF data doesn't appear when you view a photo, and you need software to view or edit it.

Geotagging was born when someone had the brilliant idea to store latitude and longitude in an unused EXIF tag so you could associate a physical location with a photo. You can manually enter GPS coordinates with a program that allows you to read and write EXIF data (one of my favorites is ExifTool, a free program available at `http://www.sno.phy.queensu.ca/~phil/exiftool`). Better yet, use a program that automatically geotags photos using saved GPS tracks. RoboGEO (`www.robogeo.com`) is a commercial product, while GeoSetter (`www.geosetter.de/en/index.html`) and GpicSync (`http://code.google.com/p/gpicsync`) are both free. Here's how these programs work.

You first make sure the clock on your digital camera is set to the same time as your GPS receiver. While you're out taking pictures, bring your GPS unit with you and turn it on so it's recording track points. When you get home, download your photos and your track log file, and then run the program. The program first checks the date and time a selected photo was taken, then it searches through the saved GPS track file for a track point that was recorded close to the photo's date and time. If a close match is found, the latitude and longitude are written to the photo's EXIF data. (A few digital cameras on the market with built-in GPS chips automatically geotag photos; Ricoh has several models. I expect more GPS cameras to appear as chip prices continue to fall.)

When a photo is geotagged, you can do all sorts of cool things with it. You can link it with Google Earth maps, share it in Flickr (which supports geotagged images), or post it at georeferenced social networking sites, such as loc.alize.us (`http://loc.alize.us`).

> ✔ **PC interface:** After you record the data, you obviously need to get it to your PC. Most data loggers have USB connectors, with some models supporting wireless Bluetooth connections. The loggers come with PC driver and interface software to set options and transfer data.

GPS data loggers tend to be cheaper than handheld GPS receivers and range in price from $50 to $100. Here are a few popular models:

> ✔ **Holux GPS Data Logger**
>
> ✔ **GISTEQ Phototracker**
>
> ✔ **GlobalSat GPS Data Logger**
>
> ✔ **Sony GPS-CS1**

Do a Google search for any of the above models, or try a more general search for *GPS data loggers,* to get more information and retailers.

GPS Radios

GPS and radios go together like peanut butter and jelly, and some nifty products on the market integrate the two. Here's a quick review.

FRS radios

FRS stands for Family Radio Service. These handheld, two-way radios are popular for outdoor recreation because they're inexpensive, don't require a license, and unlike a cellphone, are free to use (range is typically limited within several miles, depending on the terrain).

Garmin's RINO series takes FRS radios a step beyond by incorporating GPS. (RINO stands for Radio Integrated with Navigation for the Outdoors by the way, although the antenna configuration looks suspiciously like a rhinoceros horn.) Not only can you see your current position on the screen, you can also beam your coordinates to another RINO user who's in range, and your location will pop up on her map.

Garmin offers several different RINO models; you can get more information at www.garmin.com.

Ham radios

If you're into amateur radio, you should check out APRS (Automatic Packet Reporting System). APRS is a protocol that was developed in the early 1980s for broadcasting location (and other) data over ham radio frequencies. Using a compatible radio attached to a GPS receiver, coordinates are transmitted over the airwaves. Another ham radio, connected to a computer and running APRS software, receives the coordinates and maps the location. Amateur radios have considerably larger ranges than FRS radios, and APRS is often used for vehicle tracking and public safety purposes during emergencies. To learn more about APRS, visit Bob Bruninga's (the guy who developed it) Web site at http://web.ew.usna.edu/~bruninga/index.html.

Marine radios

Boaters were some of the first civilian users of GPS, and it's a little surprising it's taken so long for the marketplace to begin offering GPS-enabled radios.

Many newer marine radios incorporate something called Digital Select Calling (DSC). DSC radios have a number of nifty features, including a distress button. When you press the button, if the radio is connected to a GPS receiver, it transmits your coordinates along with a call for help. This beats just broadcasting, "Mayday, Mayday, Mayday" on your radio and having the Coast Guard trying to guess your exact position.

Make sure you know if DSC distress calls are being monitored in the waters you boat on. For example, the Coast Guard monitors DSC transmissions on coastal waters. However, getting into trouble in the middle of an inland lake and pressing the distress button won't do you much good if no one is listening.

DSC radios have mostly been the fixed types that are permanently installed in a boat. However, in 2008, both Lowrance and Standard Horizon introduced handheld DSC radios. With a built-in GPS receiver, they can send DSC distress calls as well as display your current coordinates on the screen (plus, they're waterproof and even float). Check out www.lowrance.com and www.standardhorizon.com for more information. I expect more marine electronics manufacturers to get onboard with similar products very soon.

Chapter 8

Geocaching

*P*robably the fastest way for you to get real-world experience with your handheld GPS unit is by going geocaching. This high-tech sport gets you out of the house and into the fresh air as you use both the brain in your GPS unit and the one in your head to find hidden treasure. Geocaching is a fun and challenging activity that combines modern technologies like GPS and the Internet with primitive outdoor navigation and search skills that people have been using for thousands of years. The sport also gives you a good reason to visit places you've never been.

Geocaching is pronounced *GEE*-oh-*cash*-ing. Don't pronounce cache as ca-*shay*, even if you're French. Unless you want funny looks, stick with *cash*.

This chapter is a brief introduction to geocaching. If you want to learn more, check out another book I wrote, *Geocaching For Dummies* (Wiley Publishing, Inc.).

Geocaching: The High-Tech Scavenger Hunt

When the U.S. government turned off GPS Selective Availability (SA) in May 2000, it was like magic. Suddenly civilian GPS receivers that were formerly accurate to about 300 feet were accurate to 30 feet. This level of accuracy offered some creative possibilities. Three days after SA was turned off, the following message appeared in the `sci.geo.satellite-nav` USENET newsgroup:

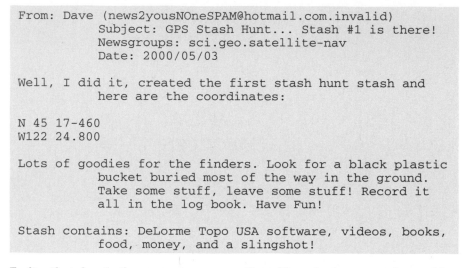

```
From: Dave (news2yousNOneSPAM@hotmail.com.invalid)
        Subject: GPS Stash Hunt... Stash #1 is there!
        Newsgroups: sci.geo.satellite-nav
        Date: 2000/05/03

Well, I did it, created the first stash hunt stash and
        here are the coordinates:

N 45 17-460
W122 24.800

Lots of goodies for the finders. Look for a black plastic
        bucket buried most of the way in the ground.
        Take some stuff, leave some stuff! Record it
        all in the log book. Have Fun!

Stash contains: DeLorme Topo USA software, videos, books,
        food, money, and a slingshot!
```

Earlier that day, in the same newsgroup, Dave Ulmer had proposed a worldwide *stash hunt,* where people would post GPS waypoints on the Internet to lead searchers to hidden goodies. While Ulmer envisioned thousands of stashes tucked in places all over the world, he had no idea how popular his idea would become.

Starting with a humble little bucket full of goodies in Oregon, Ulmer's idea took off like wildfire. Within weeks, caches were hidden in Washington, Kansas, California, New Zealand, Australia, and Chile. A newsgroup and Web site that hosted the coordinates of the stashes soon popped up as the word started to get around.

By the end of May, in a Yahoo! Group devoted to the new sport, member Matt Stum suggested that the sport be called *geocaching* in order to avoid some of the negative connotations associated with drugs and word *stash*. (A *cache* is a hidden place where goods or valuables are concealed.) Geocaching had a nice ring to it, and it didn't sound like something from a Cheech and Chong movie.

Since then, geocaching has grown very popular, and the rules are still pretty much the same: Take some stuff, leave some stuff, record it in the logbook, and have fun! Relatively cheap and accurate GPS receivers and widespread Internet access have helped the sport flourish. As of 2008, the `www.geo caching.com` site lists over half a million active caches throughout the world (compare that to a measly 72,500 sites when I wrote the first edition of this book in November 2003). That's a lot of caches out there to find! (Figure 8-1 shows a typical cache.)

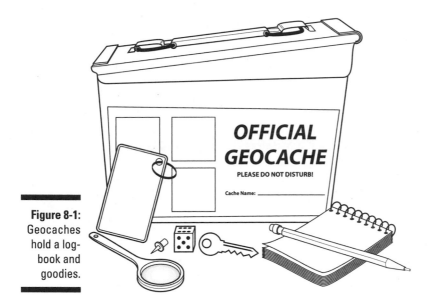

Figure 8-1:
Geocaches
hold a log-
book and
goodies.

Getting Started Geocaching

Just about anyone can participate in geocaching; gender, age, and economic status don't much matter. The main requirements are a spirit of adventure, a love of puzzles and mysteries, and a sense of fun. This section discusses what you need to geocache, how to select caches to look for, and how to find them.

What you need to geocache

Geocaching sounds pretty intriguing, doesn't it? But before you can try it out, you need a few things. You probably already have many pieces of the required gear, and your biggest investment will be a handheld GPS receiver if you don't already have one. (Read Chapter 5 for the lowdown on selecting a receiver.) With that in mind, here's a list of basic things you need.

GPS manufacturers have recognized the popularity of geocaching, and offer models with features specifically designed for the geocacher. Some newer models even support "paperless caching" that allows you to download cache locations and information directly to your handheld receiver.

Although it's possible to geocache with pocket, automotive GPS receivers, a handheld GPS is a better choice because of longer battery life, a more visible screen in direct sunlight, and outdoor-oriented navigation features.

I start with the technology-related items:

- ✔ **Cache location:** Obviously, you need to know where to look for a cache: a set of latitude and longitude or UTM (Universal Transverse Mercator) coordinates. You'll find hundreds of thousands of caches freely listed on the Internet. For information on locating a geocache, see the appropriately named section, "Selecting a cache to look for," later in this chapter.

- ✔ **Geocaching alias:** Most people who geocache use a registered *handle* (alias) instead of their real name when they sign cache logs or make Internet posts. The aliases are cool-sounding names like Navdog, Wiley Cacher, or Moun10Bike. Be imaginative and come up with an alias that fits your personality. The aliases are all unique: If you try to register an alias on one of the popular geocaching Web sites and someone else already has registered the alias, you'll need to select another name.

- ✔ **GPS receiver:** You can certainly find caches by using only a map and compass, but it's sure a lot easier when using a GPS receiver. You don't need an expensive GPS unit with lots of whistles and bells to geocache; a basic model around or under $100 will work just fine; receivers that support WAAS (Wide Area Augmentation Service, as described in Chapter 3) usually are more accurate than those that don't.

  Don't forget to bring the GPS receiver user manual, especially if you just purchased your receiver and are still trying to figure out how to use it.

A few other things can make your outing a little more enjoyable:

- ✔ **Map and compass:** A fair number of geocachers use only their GPS receiver to get them to a cache, but a good local map of the area can be very helpful. Although a receiver can lead you directly in a straight line to a cache, it's probably not going to tell you about the river, deep canyon, or cliffs between you and the cache. Even GPS receivers that display topographic maps often won't show enough detail that can help or hinder you on your way to a cache. Additionally, a map and compass serve as a backup just in case something goes wrong with your GPS. (Just make sure you know how to use them.)

Topographic maps don't provide detailed information on what the vegetation is like on the ground; therefore, I always like to look at an aerial or satellite photo of the area before I head out the door.

- ✔ **Pen or pencil and paper:** Carry a small pad of paper and a pen or pencil for taking notes about your route or things that you see on the way. Some geocachers keep an ongoing journal of their adventures, and you never know — you might turn into a geocaching Hemingway.

- ✔ **Something to leave in the cache:** When you locate a cache, you'll find all sorts of *swags,* which are treasures other people have left. Don't expect diamonds, gold bullion, or Super Bowl tickets, though. (You're far more likely to find baseball cards, costume jewelry, or corporate marketing giveaways.) Just remember that one man's trash is another man's treasure. The best things to leave in a cache are unique, out-of-the-ordinary items (such as foreign coins, fossils, exotic matchbooks, or anything that has a high cool factor). And, please, avoid leaving *McToys,* geocaching lingo for junk that you reasonably expect to find with fast-food kid's meals. For more of the lingo, check out the upcoming "GeoJargon: Speaking the lingo" section.

✓ **Appropriate clothes and footwear:** There are no geocaching fashion police, so wear clothes that are comfortable, weather appropriate, and suitable for getting dirty. Even if it's the middle of summer, it's not a bad idea to bring along a jacket in case of an unexpected rain shower or drop in the temperature. Also, make sure you're wearing sturdy and comfortable footwear if the cache is outside an urban area. High heels and wingtip loafers generally aren't recommended.

✓ **Food and water:** Some caches take all day to find, so be prepared with enough food and water to get you through your search; you can even plan a picnic lunch or dinner around your outing.

✓ **Walking stick/trekking poles:** If the terrain is really rough, a good walking stick or set of trekking poles can make life much easier when going downhill and negotiating uneven surfaces. A stick or a pole is also useful for poking around in rock cracks looking for a cache, just in case there's a creepy-crawly inside.

✓ **Digital camera:** Although definitely not a required piece of geocaching gear, a number of cachers tote along a digital camera to record their adventures or to post pictures on the Web.

✓ **Small pack:** It's much easier to put all your geocaching gear in a small daypack rather than stuffing your pockets full of stuff.

Don't forget a few safety-related items. As the Boy Scouts say, be prepared:

✓ **Flashlight:** This is a must-have for looking in cracks and crevices where a cache might be hidden — and also in case you run out of daylight. If you're smart, your flashlight uses the same type of batteries as your GPS receiver, giving you even more spare batteries.

✓ **Cellphone:** You probably have a cellphone, so bring it along (preferably with the battery fully charged). Just a note of advice, from my search and rescue experiences: I've found at times that people think of their cellphones as an absolute insurance policy against trouble. However, they can fail; cellphone batteries go dead, or you might have really bad cell coverage out in the middle of nowhere. So although a cellphone is great to have along, be prepared to take care of yourself!

✓ **Spare batteries:** I always bring along spare batteries for anything that uses them. (In this case, that means your GPS receiver and flashlight — and if you're really safety conscious, your cellphone, too.)

That's the basic gear you need for geocaching. The whole key with gear lists is to find out what works best for you. You'll probably end up carrying too much stuff in your pack at first. After you've geocached for a while, check your pack and see what you're not using so you can lighten your load.

Most geocaches are located in pretty tame, civilized areas (usually 100 feet or so off a main trail or road), but I advise letting someone know where you're going, when you'll be back, and what to do if you're late. Twisted ankles and broken-down cars seem to happen a lot in areas without cell-phone service. If you haven't spent much time in the great outdoors, check out a list of The Ten Essentials (which has been expanded to 14 over the years) at www.backpacking.net/ten-essl.html.

Selecting a cache to look for

After you get all your gear together, ready to venture out into the wilds, comes this one small detail: How do you know where to look for a geocache?

Like most other modern-day searches for information, start with the Internet. Many Web sites have listings of geocaching caches. I list some of the best Internet resources in the upcoming "Internet Geocaching Resources" section.

Geocaching.com (www.geocaching.com) is currently the most widely used site and has the largest database of geocaches all over the world; most of the information in this chapter is orientated toward that site. However, if you use another geocaching Web site, you'll find most of the same general techniques described here for selecting a cache also apply to other sites.

You can set up a free user account on Geocaching.com to view and log caches that you find, as well as to be informed of new caches that are placed in your area. Groundspeak, the company that hosts the site, also has Premium Member subscription services ($30 per year) and sells products to help defray costs.

To start, go to www.geocaching.com. Finding caches is as simple as entering the ZIP code for where you're interested in geocaching. You can also search for caches by state, city, country, latitude, longitude, or by street address (only in the United States). The basic search screen is shown in Figure 8-2.

After you enter where you'd like to search for caches, a list of geocaches in that area displays (see Figure 8-3).

The list is sorted by how far away the cache is from the search criteria you enter; the closest geocaches display first. The name and type of the cache is shown, when it was first placed, when it was found last, and how difficult the cache is to get to and find. You can scroll through the list of geocaches until you find one that looks interesting. Cache names that are lined out are no longer active.

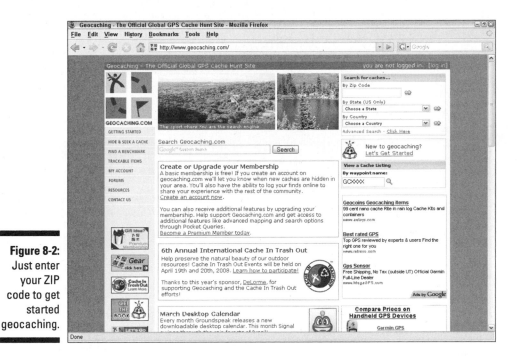

Just click the name of a cache displayed in the list about which you want more information to see a page with the following information. (Cache details displayed on the page are shown in Figure 8-4.)

- **Cache name:** The name of the cache (usually the cache name has something to do with the area where it's hidden, who hid it, or a play on words). *Bonus:* If you hide a cache, you get to name it.

- **Who placed the cache:** This is usually a cacher's alias.

- **Cache type:** Caches can be *traditional* (a single container), *multicaches* (where clues in a single cache point to one or more other caches), or *virtual* caches (a cool location that doesn't have a container).

- **Cache size:** How big the cache is (for example, a *microcache* is tiny and only contains a log).

- **Cache coordinates:** These record where the cache is located in latitude and longitude and UTM coordinates; these coordinates use the WGS 84 datum, so be sure your GPS receiver is set to this datum.

- **When the cache was hidden:** The date the cache was originally placed.

- **Cache waypoint name:** All caches in the Geocaching.com database have a unique name: for example, *GC* followed by the numeric order the cache was added to the database. You can use this to name a waypoint on your GPS receiver for the cache location.

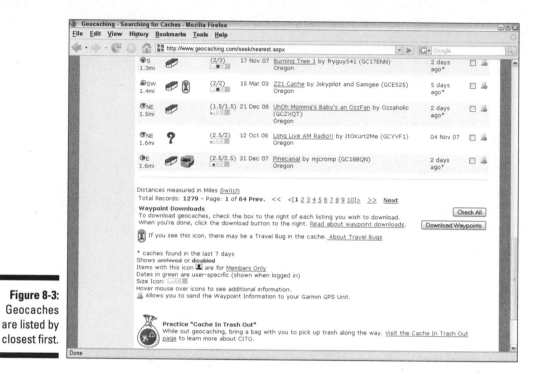

Figure 8-3:
Geocaches
are listed by
closest first.

✔ **Difficulty:** The difficulty rating is how hard the cache placer thinks the cache will be to find; 1 is easiest, and 5 is the most difficult. Whoever places the cache decides the difficulty level, based on some general criteria, such as how steep or rocky the terrain is or if you have to go through very much underbrush to reach the cache.

✔ **Terrain:** The terrain rating is how difficult the terrain is: 1 is flat, easy, and level; 5 could be very steep and rocky with lots of underbrush and generally miserable travel conditions. As with the difficulty rating, it's up to the cache placer to rate the terrain.

✔ **General description of the cache:** Cache descriptions range from a couple of sentences to stories and history lessons about the location. Clues often appear in the description, so check it closely.

✔ **Map location of the cache:** To the right of the page is a map that gives you a general idea of where the cache is. A larger map with more detail appears at the bottom of the page. You can click the larger map and go to the Yahoo! Maps Web site, where you can zoom in on the cache site.

✔ **Hints:** The cache placer can optionally add hints to help a geocacher narrow his search. The hints are in code; I discuss these in the following "Finding the cache" section.

✔ **Logged visits:** A list of all the comments about the cache from people who have visited it and then logged Web site comments.

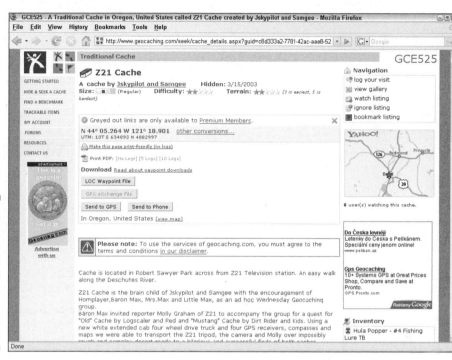

Figure 8-4: Information about a cache includes its name, its location, and a description.

Some of these logged visit comments may contain *spoilers,* which are hints that may make it easier to find the cache. Although most cachers try not to spoil the fun for others, sometimes a clue accidentally appears.

Before heading out to search for a cache, check the last time someone found it. Although Geocaching.com tries to keep track of inactive caches, sometimes caches that have been stolen or kidnapped by space aliens slip through the cracks. If you're just getting started geocaching, go after caches that have had some recent activity. This increases the odds that they'll still be hiding where they're supposed to be when you go looking for them.

Finding the cache

After you select a cache you want to search for, the next step is finding it. This might be a little bit more challenging than you think. Remember that your GPS receiver will only get you within 10–30 feet of the cache location — perhaps even farther away if you have poor satellite coverage or the cache hider's coordinates are a little off. After your GPS unit gets you to the general vicinity of the cache, start using your eyes and your brain, which at times might be more reliable than your GPS receiver.

Sometimes a series of caches are located close together, usually separated by at least a tenth of a mile. Because you're already in the neighborhood, consider trying to find several instead of going for just a single cache. Check out the link on the cache description page that displays all the nearby caches and how far away they are from each other.

Finding a cache boils down to following these general steps (be sure to read the "Geocaching Etiquette" section later in the chapter):

1. **Enter the cache coordinates in your GPS receiver as a waypoint and give it a name.**

   The methodology for entering waypoints differs from model to model. Check your user manual for specific instructions about how to enter and name a waypoint on your GPS model. Read more about this in Chapter 4.

   You can use the six-character waypoint name on the cache description Web page for the name of the waypoint. Double-check that you've entered the correct coordinates in your GPS receiver. Many caches aren't found on the first try because of a typo in the coordinates.

   If you have an account on Geocaching.com, you can download the cache waypoint to your computer from a link on the cache description page and then upload the waypoint directly to your GPS receiver. Doing so helps to eliminate errors caused by boo-boos with the GPS coordinates.

2. **Print a copy of the cache description Web page so you can bring all the information you need to find the cache with you.**

   If your printer is out of ink or you're being frugal, scribble down the coordinates and any other information that you think might be useful in locating the cache.

3. **Gather your equipment, including your GPS receiver, map, compass, food, water, and other essential items mentioned in this chapter.**

4. **Head out to the cache's starting point.**

   Drive or bike as close to the cache as you can get. Sometimes the cache descriptions give you exact instructions, like at which parking lot or trailhead to start from. The more challenging caches give you only the coordinates; it's up to you to decide where you'll begin and how you'll get there. One of the pleasures of geocaching is it's usually not a timed event (although some timed competitions are out there), and you can take as long as you want to reach the cache site, stopping to smell the roses and enjoy interesting sights.

5. **Turn on your GPS receiver and get a satellite lock.**

   Hopefully! If not, you brought that map and compass, right?

6. **Save a waypoint for your starting point.**

Getting back to your car can sometimes be a challenge after finding a cache, and saving a waypoint with your car's location can make life much easier (and get you home in time for dinner). Your GPS manual contains details for setting a waypoint for your particular model.

7. **Double-check to make sure that you have the coordinates, cache description, hints, and the rest of your geocaching equipment in your possession. (Keeping it all together in a backpack is convenient.)**

   From personal experience, I can tell you it's never any fun arriving near the cache and finding I left the cache description, with vital clues, back in my car that's now a couple of miles away.

8. **Activate the cache's waypoint.**

   Activating a waypoint tells the receiver to calculate the distance and direction from your current spot to the waypoint's location. Your GPS unit will let you know how far away the cache is and what direction you need to head to get there. (This often is as simple as pressing a button on the GPS receiver and selecting the waypoint you want to go to.)

9. **Follow the direction arrow, road map display, or compass ring on your GPS receiver toward the cache.**

   A local map can come in handy as you move toward the cache because you can use it to figure out what the terrain is like and whether any rivers, cliffs, or mountains lie between you and the cache. You can use some of the online digital maps discussed in Chapter 18 for preplanning or print them out to use in the field.

   Don't feel compelled to always head in the direction your GPS unit tells you to go. It might make more sense to walk around a pile of rocks or downed trees than to go over the top of them. After you get around an obstacle, you can check your receiver again to get on the right course.

   Watch your step! As you head toward the cache, don't get so caught up in staring at your GPS receiver that you fall off a cliff or trip over a tree root. And watch the scenery, too. Usually the journey is the reward.

10. **When your receiver says you're within 30 feet or so of the cache, move around and find the place that reports the closest distance to the cache.**

    Begin your search at that spot. This is where the real fun starts. You now shift from relying on technology to using your powers of observation and common sense. A cache could be inside a cave, tucked in a tree hollow, hiding behind a rock outcropping, or concealed under a pile of brush. Some caches are easy to find, and others are devilishly difficult.

When you start looking around, you can do a few things to help improve your odds of finding the cache:

✔ **Find out the maximum distance to the cache.** Check the Estimated Position Error (EPE) to see how accurate your GPS receiver currently is, based on the satellite coverage. *Remember:* The bigger the number, the less accuracy. This helps you roughly determine how large your search area is. For example, if the EPE is 20 feet, your search area is a circle with a 40-foot diameter, with the center at the closest location that you can get to the cache waypoint.

✔ **Follow a magnetic compass.** When you're within 30 feet of the waypoint and your GPS receiver is showing a consistent bearing to the cache (tree cover and poor satellite coverage can cause the distance and direction numbers to jump around), use a magnetic compass to guide yourself to the cache. As you slow down, unless your receiver has an electronic compass, the direction that your receiver shows to the waypoint becomes less precise, and you can easily veer off-course. Handheld magnetic compasses or electronic compasses built into the GPS unit don't rely on satellite signals, and won't have this problem.

✔ **Think about the container.** Knowing what kind of container the cache is stored in can be a big help in identifying and eliminating possible hiding spots. Sometimes the cache description lists the container type (ammo can, plastic ware, bucket, or whatever), which can narrow your search based on the container size and shape. For example, you shouldn't be looking for an ammo can in a 3-inch-wide crack in a rock.

✔ **Think about the terrain.** Look at the surrounding environment to get a general idea of where a cache might be hidden. What natural (or man-made) features make a good hiding place? Remember, unlike pirate booty hiding, geocaching has a rule against burying cache containers, so you shouldn't be burrowing holes like a gopher.

✔ **Split up the work.** If you're geocaching with other folks, assign areas for people to check. Although you don't need to precisely measure and grid-off squares, divvying up an area to search is faster and more efficient than randomly wandering around.

✔ **Think like a cache hider.** If you were going to hide a cache, where would you hide it? Sometimes trusting your intuition can be more effective than trying to apply logic. After you check the ordinary places, start looking in the unordinary spots.

There's an old safety saying in wildland firefighting that goes, "Look up, look down, look all around." The same advice applies to geocaching, which is an excellent way to improve your overall awareness and observation skills.

Going in circles

This will happen: Your GPS unit tells you you're in the general vicinity of the cache, but after a couple of hours of wandering around in circles, you still can't find the cache container. You've double-checked the coordinates, the satellite coverage is good, and you're starting to get a bit frustrated. Take a deep breath. Here's what to do.

You can always resort to using a hint. Most cache description pages have a short hint, but you have to work for it because it's encoded. The reason for the spy stuff is so the hint doesn't spoil the fun for another cacher who doesn't want to use the hint as part of his search. Fortunately, the hint is encoded with a simple substitution code (for example, A = M, B = N, C = O, and so on), so you don't need to work for the NSA to be able to break it. The decoding key is on the right side of the page, and it's pretty easy to figure it out by hand. See Figure 8-5 for what a coded hint and decoding key look like.

Figure 8-5: If you're stumped (ha!), go for a hint.

One of the more challenging types of caches to find is a *microcache*. Instead of using large containers, smaller ones — like pill bottles, 35mm film canisters, or magnetic hide-a-keys — are used that only hold a piece of paper that serves as the log. These caches are typically in urban areas and are cleverly hidden to avoid detection by nongeocachers passing by.

In addition to the hint, you can also look through the logged visit comments that other people have posted who have already found the cache. Although most geocachers try to avoid including *spoilers* (way-too-obvious hints or commentary) into their comments, sometimes enough information leaks through that can help you narrow your search.

How you go about finding the cache is up to you. Some purists will use only the coordinates and basic description of the cache, never using the hints or the comments. Other cachers decrypt the hint and read all the comments before they head out the door on a search. It's your choice.

There's no shame in a DNF (Did Not Find); it happens to everyone. Go back to the cache location another day and try again. Geocaching is supposed to be fun, so don't take it too hard if you can't locate a cache. Consider bringing someone else with you next time: Two heads are better than one, and a different set of eyes might find something you overlooked. Don't be shy about logging a DNF for the cache at the Geocaching.com site. If a cache owner hasn't visited the site in a while, a number of logged DNFs could mean that the cache has been moved or stolen by someone. Unfortunately, cache vandalism and thievery happen: For example, the cache you were looking for might have been stolen, and the database hasn't been updated yet.

Lost but now found

In more cases than not, after looking around for a while, behold! You find an old, olive-drab ammo can tucked behind some rocks. Congratulations, you found your first geocache! Now what?

- **Savor the moment.** There's definitely a sense of accomplishment when you find a cache and a little bit of childlike wonder as you open up the container to see what types of treasures are inside.

- **Sign the logbook.** Write down the date, a few sentences about your experiences finding the cache, what you took and/or added, and your geocaching alias. Some people who are really into geocaching have custom business cards or stickers made up for placing in the cache log.

- **Read the logbook.** It's fun to read about other cachers' adventures and when they discovered the cache.

- **Exchange treasures.** If you take something from the cache, leave something. If you forgot your goodies, just sign the logbook. Quite a few geocachers are more into the hunt for the cache than for the loot inside.

  Trading up means leaving something in the cache that's better than what you take. There's always been a considerable amount of discussion in the geocaching community about how caches start out with cool stuff but soon end up filled with junk (broken toys, beat-up golf balls, cheap party favors, and so on.). Some geocachers even take it upon themselves to remove anything from a cache that doesn't meet their personal quality bar. If you can, trade up to make the finds more interesting for everyone.

- **Cover your tracks.** Seal the cache container up (including the sealable plastic bag that usually contains the logbook) and put it back where you found it, making sure that it's hidden just as well as it was when you arrived.

✔ **Go home.** Use the track-back feature of your GPS receiver to follow your exact path back to your car. Better yet, activate the waypoint that you set for your car but see some different sights by taking a new route back to where you started.

✔ **Share your experiences.** When you get back to your computer (if you're a member of Geocaching.com), log your find on the Web site so the whole world knows you found the cache. Go to the cache description page and click the Log Your Visit link at the top of the page. (This is completely optional. Some geocachers prefer operating under a low profile, keeping their discoveries and adventures to themselves.)

✔ **Do it again (and again, and . . .).** After you have your first cache find under your belt, you're ready to venture out again into the brave new world of geocaching and find even more caches. As your experience with a GPS receiver grows and your skills in navigation and cache finding improve, you'll likely want to start challenging yourself more by going after caches that are more difficult to find and reach.

GeoJargon: Speaking the lingo

Like any sport or pastime, geocaching has its own language. Because the sport is still relatively new, the jargon is evolving, but here are some terms to be familiar with so when you talk to other people about geocaching, you sound like a pro.

✔ **Archived:** Caches that no longer exist but still appear in a Web site database for historical purposes. A cache can be archived because it's been stolen, is no longer maintained, or doesn't abide by the guidelines for where caches should be placed.

Geocaching stats

Just like any sport, geocaching has statistics (stats). In this case, *stats* refer to the number of caches that you've found and hidden. When you sign up for a free or premium account at Geocaching.com, you can log the caches you've found as well as add caches that you've hidden to the site's extensive database. The Web site tracks the finds and hides for you and displays them on a user profile page. Other members can check out your stats, and the number of caches that you've found appears next to your alias when you log your comments about a cache you've visited. Some geocachers are competitive and are in to racking up as many cache finds as possible. Others are more blasé about the whole numbers thing and could care less. Like so many other aspects of geocaching, it's up to you how you want to play the game.

- **Cache machine:** A preplanned event in a local area, where geocachers look for caches. The event can last hours or days. This is a marathon-endurance session of geocaching, where you try to find as many caches as you can in a set amount of time. The event is named after a dedicated geocacher. BruceS, a true cache machine, found 28 caches in 24 hours, totaling 86 finds in 5 days.

- **DNF:** Did Not Find (as in, *did not find the cache*). It happens to everyone. If you didn't find the cache, try again on another day.

- **FTF:** First to Find. This means bragging rights that you were the first person to find a newly placed cache.

- **GPSR/GPSr:** GPS receiver. Many people drop the R and just call a GPS receiver a *GPS*.

- **Hitchhiker:** An object that moves from cache to cache. A hitchhiker is marked with some instructions, telling the finding geocacher to take it and place it in another cache.

- **McToys:** Cheap trinkets left in a cache, like the toys that appear in fast-food kids' meals. There are better things to leave in caches.

- **Muggles:** People you encounter on the trail who aren't geocachers; from the Harry Potter stories.

- **Neocacher:** An inexperienced or newbie geocacher.

- **Signature item:** Something unique that a particular geocacher always places in a cache that he or she finds.

- **Spoiler:** Information that might give away the location of a cache.

- **Swag:** Goodies that you find in a cache; from the marketing term *swag* (or schwag) used to describe the promotional trash and trinkets (tchotchkes) handed out at trade shows.

- **TNLN:** Took Nothing, Left Nothing. Just what it sounds like. Also, *TNLNSL,* which means that the geocacher signed the cache log.

- **Travel Bug (TB):** A type of hitchhiker that you mark with a special dog tag purchased from Geocaching.com. When TBs are found, their journey is tracked on the Geocaching.com Web site. Travel Bugs can have specific goals (as in, getting from Point A to Point B) or are just released into the world to see how far they can travel.

Hiding a Cache

After a while, you might get the urge to set up a cache of your own. This section discusses how to create and hide a cache. It's not that difficult, and most cache hiders spend $10 or less to set up their cache, which is some pretty cheap entertainment these days. It's also a way to give something back to the sport.

The force is strong with this one

A *Star Wars* Darth Vader action figure, Travel Bug, started his travels in an Arkansas geocache in February 2002. By the time he was returned to his owner nine months later, he had journeyed 17,534.64 miles, including tagging along on aerial missions in Afghanistan, pub-hopping in England, and working on his tan in Florida. At last report, Darth was getting some R&R in Texas.

 Don't rush out and hide a cache before spending some time finding caches. Searching for other caches will give you some good ideas and set expectations for creating your own. Check out Geocaching.com for FAQs as well as a complete set of guidelines for hiding and placing caches.

Selecting a container

First things first. You need something to house your cache in. The only real requirement for the container is that it needs to be waterproof, although sometimes cachers use plastic bags inside a non-waterproof container. The size of the container determines where you'll be able to hide the cache and how full of trading trinkets it can become. Any container that you can think of has probably been used for geocaching, including plastic buckets with lids, breath mint tins, margarine tubs, 35mm film canisters, pill bottles, plastic Army decontamination kit boxes, and PVC piping. You probably have a suitable geocaching container lying around the house or garage. Just for the record, the two most popular types of cache containers are

- **Ammo cans:** Made of military surplus steel, ammunition (ammo) cans work great because they're sturdy and waterproof. They typically come in two sizes, based on the machine gun ammunition they once held:

  - *50 caliber:* 11 inches long, 5.5 inches wide, and 7–5 inches deep

  - *30 caliber:* 10 inches long, 3.5 inches wide, and 6.75 inches deep (The narrow cans tend to fill up with trinkets quicker.)

  Depending on the terrain and vegetation, the olive-drab color makes ammo cans difficult to spot. You can typically get ammo cans for around five dollars or less from local or online Army surplus stores.

- **Tupperware:** Rectangular Tupperware or other plastic storage containers are also a popular choice but aren't quite as rugged as an ammo can. Sometimes a geocacher doesn't reseal the lid very well. Plastic containers are cheaper and more available than ammo cans, and you can easily match a size to go with any cache. Some cache hiders spray paint their containers to make them blend better with the surroundings.

Location is everything

Just like in real estate or retail sales, location is everything when it comes to placing a cache. After you select a container, figure out where to put it. The location of your cache usually defines its success and popularity.

I recommend doing some initial research first to locate a general area to hide your cache. For many geocachers, visiting a new place with some unique feature, incredible scenery, or just gorgeous view is every bit as important as finding a cache. Keep this in mind as you use maps, travel guides, or memories from your own explorations to help you select a good cache location.

An important part of your homework is discovering where caches are and are not permitted. The geocaching community tends to be very aware that the continued growth and success of the sport depends on good relationships with landowners.

If you want to place a cache on private property, always first ask the owner's permission. Because not everyone knows about the popular, somewhat geeky pastime, take the time to explain how the sport works.

Always check with a governmental agency before placing caches on its land. You can contact the agency directly, try a Google search to see whether its geocaching policies are published on the Web, or talk with other geocachers in your area to get their experiences in dealing with different agencies. For example, the U.S. Bureau of Land Management recognizes geocaching as a recreational activity and tends to be friendly toward cache hiders who want to locate a cache in places other than wilderness or wilderness study areas. The U.S. National Park Service, on the other hand, prohibits placing geocaches on the land that it manages; if you're caught hiding a cache on such land, it's a federal offense. Yipes.

After you figure out the land ownership issues, the next step is to ensure that your cache appears in the Geocaching.com database. The site has a series of common-sense criteria that a cache must meet to be added to its database. Generally a cache can't be

- **Buried:** Covering it with braches, rocks, or leaves is okay, but no digging, please.
- **Placed in environmentally sensitive areas:** This includes archaeological and historic sites.
- **Placed in national parks or designated wilderness areas:** This is a no-no. Sorry, them's the rules.
- **Placed within 150 feet of railroad tracks.** Umm, this is for safety reasons.

- **Placed anywhere that might cause concerns about possible terrorist activities:** Use your post-9/11 brain. This includes areas near airports, tunnels, military facilities, municipal water supplies, and government buildings or bridges. Don't risk an all-expenses paid trip to Guantanamo.

- **Placed within one-tenth of a mile of another cache:** This is a rule for adding a cache to the Geocaching.com database as well as simple geocaching etiquette.

- **Of a commercial, political, or religious nature:** Keep it neutral and tasteful (this is a family sport); don't cache something promoting some business or cause.

The geocaching community polices itself fairly well. If you put a cache where it shouldn't be, a cacher will probably let the Geocaching.com administrators know about it, and the cache will be removed from the database.

After you select a good general location to put the cache, visit the area to figure out exactly where you're going to hide the cache. Use your creativity to find a challenging hiding place: in a tree hollow, underneath bushes, wedged in rocks, and so on. The more experience you have finding caches, the more ideas you'll have for good hiding places.

After you find your secret hiding spot, you need to determine the cache's coordinates as precisely as possible. (Use the WGS 84 datum; see Chapter 4 for more on this.) This can be challenging because of less-than-perfect satellite coverage. You might find the location's coordinates changing on your GPS receiver every few seconds. Many GPS units have an averaging feature that compares coordinates at a single spot over a period of time and then averages the result. If your receiver does do averaging, get it as close to the cache as possible, let it sit for five or ten minutes, and then copy down the cache coordinates and enter them as a waypoint.

A manual approach to averaging is to set a waypoint for the cache location, walk away, and then come back and set another waypoint. Repeat this until you have 6–12 dozen waypoints; then examine the list of waypoints, and pick the one that looks the most accurate (generally the value in the middle of the list).

Stocking the cache

Here are the basics of what to cache in your cache:

- **Logbook and writing utensil**

  At the very minimum, your cache should contain a logbook and a pen or pencil so other cachers can write about their discovery. (Pencils work better in cold climates because the ink in most pens will freeze; mechanical pencils are the best because they don't need sharpening.) The logbook is usually a spiral notebook with the name of the cache written on the cover. Some cache hiders paste their personal logo or some other graphic to the notebook cover. As the cache founder, you should write some profound thoughts about the cache on the first page.

✔ **Identifying information**

The cache should have some information that identifies it as a geocache, describes what geocaching is, and provides instructions to the finder. (Just in case non-geocachers stumble upon the cache.) The Geocaching. com site has an information sheet in a number of different languages that you can print out and place in your cache; laminating this sheet is a good idea. Be sure to mention the cache's name and its coordinates.

✔ **Treasures**

Add some treasures to your cache. These should be unique and interesting items. Because geocaching is a family sport, initially put a mix of things in it that appeal to both adults and children. You don't need to fill the container up like a stocking at Christmas. Many caches start out with 6–12 small items. If you want, you can add a hitchhiker or a Travel Bug. (Read more about both of these critters in the earlier section, "GeoJargon: Speaking the lingo.")

Even though your storage container may be waterproof, always put your logbook and cache goodies into resealable plastic storage bags. This prevents your cache from turning into a soggy mush when someone inevitably forgets to seal the container's lid.

Submitting the cache

Time for a little advertising: It doesn't do much good if people don't know about your cache after you place it. The Geocaching.com Web site currently has the largest database of caches and is where most people go to find information about caches. You need to have a free or premium account at the site to be able to post your cache, so if you don't have an account yet, go to the site and sign up. (I promise that it's quick and painless.)

After you log on to the site, submitting a cache is just a matter of filling out an online form about your new pride and joy. You enter things like the cache's name (think of something creative), its coordinates, the date it was placed, and other information similar to what you find when you're looking at a cache description Web page.

If you're having trouble trying to determine the terrain and difficulty ratings, head over to geocacher ClayJar's online terrain and difficulty calculator at www.clayjar.com/gcrs.

Taking geocaching to the extreme

Although geocaching usually doesn't require a high degree of fitness or special skills, a few caches out there might be labeled *extreme geocaching*. A cache might be perched midway down a cliff face, requiring climbing equipment to rappel down to reach it. And handfuls of caches are underwater and can only be reached by scuba diving. (GPS doesn't work underwater, so this would be the spot for a boat to anchor.) Obviously, these types of caches limit the number of finders but can be quite unique if you're into challenging and technical outdoor sport.

After you enter all the cache information, submit the form. Volunteers will check things such as whether all the information needed is present, the coordinates are generally correct, and the cache meets the general submission guidelines. Keep in mind that volunteers don't physically visit the cache because that would require thousands of people all over the world with a considerable amount of free time on their hands. The approval process can take up to a couple of days but is usually shorter. If you're approved, your cache is added to the database. If you're not approved, you'll be informed why, and you can either address the problem and resubmit or discuss the issue with the staff.

If you're handy with HTML and your Internet provider supports Web hosting (or you use a free Web page hosting service like Google Pages or MySpace), you can associate a Web site with your cache. The Web site might have digital photos, detailed maps, or anything else that supplements or complements the standard information found in a cache database entry. You then list the link in the cache description.

Maintaining the cache

After you hide your cache and it appears in the database, your work isn't over yet. You now have the responsibility of maintaining the cache. This means visiting the cache every now and then to verify that it's still there and in a good state of repair. You may even need to restock it with some trinkets if the supply is running low. During your visits, check that the area around the cache isn't being extremely impacted by people searching for the cache. If the site is being disturbed, consider either moving the cache to a new location or pulling the cache entirely. (If you decide to temporarily or permanently remove a cache, be sure to post a log entry to let other geocachers know when they look up information about the cache. Also, let the Geocaching.com administrators know so they can update their database.)

In addition to physically checking the cache, you should also check your cache online and read the comments posted from people who have visited the cache. These comments can alert you when it's time to make a maintenance call to the cache. Patience, Grasshopper! Sometimes it can take a while for someone to first find your cache and post about it.

Geocaching Etiquette

For the most part, there aren't a whole lot of rules when it comes to geocaching. It mostly boils down to respecting other cachers and the land that you play on. Consider these etiquette points when you're out geocaching:

- **Always respect private property.** Need I say more?

- **Don't leave food in a cache.** Food can attract animals as well as get smelly and messy, and plastic cache containers have been chewed through to get at a tasty snack.

- **Never put anything illegal, dangerous, or possibly offensive in a cache.** Again, geocaching is a family sport, so be responsible.

- **Always trade up or replace an item in the cache with something of equal value.** Don't be a Scrooge; what's the fun in that?

- **Be environmentally conscious when searching for and hiding caches.** Tread lightly on the land. Check out the Leave No Trace site at www. lnt.org for more information.

- **Geocaching is a pretty dog-friendly sport.** Keep it that way by having Fido tethered in leash-only areas. And no matter how good your dog is, have a leash ready in case other people or animals are around.

- **Cache In, Trash Out (CITO).** If you see any litter on your way to or from a cache, get some additional exercise with a deep-knee bend, pick it up, and pack it out.

- **Say thank you.** After you visit a cache, send a quick e-mail, thank-you message to the geocacher that placed the cache or acknowledge him in the online cache comments.

Internet Geocaching Resources

Because geocaching is very much a sport of the Internet community, the Internet contains some terrific sites about the sport. Here's a sample:

Cache hider's checklist

When you hide caches, you bring along most of the same things you have when you search for caches. Here are a few other things not to forget:

✓ Waterproof cache container

✓ Cache log (spiral notebook)

✓ Pencils and/or pens to leave in the cache

✓ Resealable plastic bags

✓ Trinkets to stock the cache

✓ Notebook to record information about the cache to submit to the Geocaching.com database

---

✓ **Geocaching.com** (www.geocaching.com): This is the primary geocaching site on the 'Net. In addition to an extensive database of caches and FAQs about the sport, the site also has a large number of forums dedicated to different geocaching and GPS topics.

✓ **Navicache.com** (www.navicache.com): This is the second-largest Web site dedicated to geocaching, but it's still currently quite a bit smaller than Geocaching.com in terms of caches listed. The site has many of the same features as Geocaching.com and is often viewed as an alternative to the more mainstream, larger site. There's not much duplication in the cache listings between the two big sites, so be sure to check both their listings when searching for caches.

✓ **TerraCaching.com** (www.terracaching.com): A members-only geocaching site that prides itself on high-quality caches and encouraging competition among not only the cache seekers but the cache hiders too.

✓ **Buxley's Geocaching Waypoint** (http://brillig.com/geocaching): This site has a comprehensive set of maps that provides a bird's-eye view of non-Groundspeak caches in your area. Just click a dot on the map for cache information.

If you want to socialize with other geocachers in your area, local and regional clubs and Web sites have sprung up. Many of these sites have their own lists of caches and practical information for the novice or experienced cacher. Do a Google search for *geocaching* and your city or state to search for Web sites with more information. You can also find out more about geocaching in your area by checking out the regional discussion forums at Geocaching.com.

The geocaching community is not immune to politics. Skirmishes and large-scale battles can break out between individuals and rival Web sites. It's best to duck your head, check your GPS receiver, and just head to the cache waypoint.

Other caching pursuits

In addition to geocaching, a number of other GPS-related activities have sprung up on the Internet. A few that you might be interested in include

✔ **Geodashing:** A contest in which random points are selected and players need to get within 100 meters of the location. There are no caches, hints, or terrain difficulty ratings, and the points can be anywhere on Earth. In fact, some locations can be impossible to reach. A new contest takes place roughly every month. The goal of the game is for teams to collect all the points first or to get as many as they can before the contest ends. For more details, check out http://geodashing.gps games.org.

✔ **The Degree Confluence Project:** An interesting project in which people use their GPS receivers to visit places where latitude and longitude lines converge. They take a digital picture, which is then published on a Web site. The goal is to map all the major latitude/longitude intersections for the entire Earth. For more information, go to www.confluence.org.

✔ **Benchmark hunting:** *Benchmarks* are permanent markers installed by the government for survey purposes. Over one-half million benchmarks have been installed in the United States. The most familiar type is a small, brass disk embedded into rock or concrete. The National Geodetic Survey (www.ngs.noaa.gov) maintains a list of the benchmarks and their locations. The Geocaching.com site also provides benchmark locations and lets you log a benchmark when you find one.

✔ **Whereigo:** Whereigo is a series of new, GPS-enabled adventure games from the people who run the Geocaching.com Web site. They've developed a toolkit that allows programmers to build adventure games that rely on GPS receivers. It's an interesting idea, but the concept only works with GPS PDAs and certain Garmin GPS receiver models (there are plans for the game to be available on mobile phones and other GPS units). Time will tell whether Whereigo will become as popular as geocaching. For more information, go to www.whereigo.com.

✔ **Waymarking:** A sister site of Geocaching.com where interesting places all over the world are cataloged (including GPS coordinates, photographs, and descriptions). There's no hidden treasure, only the experience of visiting (and logging, if you like) some place new. To check it out, visit www.waymarking.com.

✔ **EarthCache:** A wonderful educational resource where instead of looking for treasure you learn about natural features of the earth, such as rivers, volcanoes, faults, and fossils, up close and personal. Photos and descriptive information is included along with GPS coordinates. The Geological Society of America and various commercial, government, and non-profit organizations sponsor the site. Visit it at www.earthcache.org.

✔ **GPS Drawing:** This is an interesting form of art based on using your GPS receiver to record where you've been. For some amazing examples, check out the gallery at www.gpsdrawing.com.

Part III
Digital Mapping on Your Computer

The 5th Wave   By Rich Tennant

"Okay, Darryl, I'm pretty sure we didn't load the digital mapping software correctly."

In this part . . .

Time to get practical. This part is all about digital maps on your PC — desktop mapping so to speak. You find out about hardware (both basic requirements for desktop mapping and connecting a GPS receiver to a PC) and review a number of different software packages you can use to access aerial photos and topographic and street maps.

Chapter 9

Digital Mapping Hardware Considerations

· ·

In This Chapter

▶ Determining hardware requirements for different mapping activities

▶ Weighing hardware and peripheral digital mapping considerations

· ·

*I*f you're planning to use digital mapping software on your PC, you might wonder about computer hardware and peripheral requirements (like a printer). What kinds of computer and add-ons work best for creating and using digital maps?

Your answer depends on what type of mapping you'll be doing and what kind of software you plan to use to access and create maps. In this chapter, I lead you through your choices to give you a pretty good idea whether your current computer can meet your mapping needs or whether you need to think about investing in a new machine. I also cover how much storage you might need as well as what makes a good printer for mapmaking.

An Internet connection is a must for anyone interested in digital mapping.

Digital Mapping Software Choices

Understandably, your hardware needs are driven by your software requirements. Here are the three main types of mapping programs you're most likely to use:

✔ **Commercial mapping programs:** Commercial mapping programs come bundled with maps and offer a number of powerful features but are relatively easy to use. Most commercial map programs don't have extensive hardware requirements. In fact, many of the programs on the market work fine with older computers. (I discuss GPS receiver manufacturer software in Chapter 11 and street and topographic map software in Chapters 13 and 14.)

✔ **Web-hosted map services:** Web-hosted map services are accessed with your Web browser. These map Web sites are easy to use but don't offer as many features as commercial or standalone mapping programs. Viewing Web site maps isn't a very resource-intensive activity. The speed of your Internet connection is a bigger issue than the speed of your computer's processor. (For more on using Web-hosted map services of various types, see Chapters 18 and 19.)

✔ **Standalone mapping programs:** Standalone map programs are similar to commercial map software but don't come bundled with maps; you need to provide the map data yourself. If you're using a standalone program to make maps from data that you download from the Internet — especially if you'll be creating 3-D images — you want as much processor speed, memory, and storage as your budget allows. It also doesn't hurt to have a high-speed Internet connection to speed up downloads. (I talk about some standalone map programs in Chapters 14, 15, and 16.)

Additionally, you need to factor in your processing power as well as what types of storage devices, display devices, printers, and communications equipment you'll need and use.

In most cases, if you've purchased a computer in the past couple of years that runs Windows, it's probably going to be more than adequate for computer mapping. (Always check a mapping software vendor's Web site first, though, to ensure that your computer is compatible with the program you plan to use.)

Processing Power

When it comes to software, whether it's an operating system or program, the processor and amount of memory your PC has can make a big difference in performance. Some people think that just like money, you can never have enough processor speed or memory — when it comes to mapping software, though, that's not entirely true.

You don't absolutely need to have the latest cutting edge and fastest technology for computer mapping. You can easily get by using older computers and peripherals. I still have an old 500 MHz Pentium with 256K of RAM (shudder, when dinosaurs ruled the earth) running Windows 2000 that works fine for basic mapping. Granted, it's slower performing some tasks compared to faster, newer computers, but it still gets the job done. With this in mind, take a look at practical processing and memory requirements for mapping software.

If you plan on going mobile with your maps, laptops offer obvious advantages over desktop PCs. Mini-laptops such as the Asus Eee PC are especially well suited for loading up with map software.

Processors

Most commercial mapping programs have pretty humble processor requirements. Based on the manufacturer system requirements, to use some of these programs, your computer should have a modern Pentium or similar chip with a minimum speed of 300 MHz (I think my cellphone might have a faster chip than that). That's a pretty meager amount of computing power considering that current computers offer at least seven times greater processor speed, if not more. If you're using commercial mapping software, just about any contemporary computer is going to fit the bill when it comes to processor requirements.

You'll want a faster processor and more computing horsepower if you're doing a lot of 3-D mapping or processing large amounts of map data — particularly creating maps from data that you download from the Internet. This can be a processor-intensive task: The faster the chip, the quicker the map or terrain image will be rendered and displayed. (Any Pentium III or above PC with a processor speed of over 1.2 GHz should suffice for the average map user.)

Memory

If you look at the system requirements of commercial mapping software, you'll see some ridiculously low memory requirements considering what's standard in today's computers. (When was the last time you saw a computer advertised with 16, 32, or 64K of RAM? That's actually some of the stated minimum memory requirements for a few popular mapping programs.) Every contemporary computer should have enough memory to work with most mapping programs.

A mapping program may have miniscule memory requirements, but don't forget about the operating system. If you're running Microsoft Windows XP, you should consider running at least 512K of RAM. Vista needs, at minimum, a gigabyte of RAM. (Speaking of Vista, be sure to check if your mapping software is compatible with the operating system.) With modern operating systems, memory is one of those things you can't have enough of.

Don't fret over the different types of memory. Double Data Rate (DDR) memory chips are indeed faster than conventional, synchronous dynamic RAM (SDRAM) memory. However, if you're the average computer user doing typical mapping projects, you're probably not going to notice the difference.

Storage Capacity

Another important hardware consideration is storage capacity. Map data files don't tend to be small like word processing or spreadsheet documents, so you need to plan accordingly. This section gives you some advice on ensuring that your storage space is adequate for your maps.

Hard drives

In these days of cheap, large hard drives, it's easy to get a little blasé about storage space. Digital mapping can take up quite a lot of hard drive space, though, and you should be aware how much space your map software and its data could consume.

Software storage needs

A mapping program can easily install between 300–500MB on your hard drive, and that doesn't count all the map data that's contained on a CD or DVD. Depending on program options and the types of maps to be used, you can easily have several gigabytes of space taken up by a single mapping program. Always check the software's hardware requirements to ensure you'll have enough storage space to install the program.

If you're running low on hard drive space, some mapping programs have a minimal install option that leaves some program data on the CD or DVD instead of writing everything to the hard drive by default.

Data storage needs

Most commercial mapping programs come with map data on CDs or DVDs, so you shouldn't need to worry about storage space unless you plan to copy the map data to your hard drive.

However, if you're downloading lots of raw data from the Internet to create your own maps, you definitely need to think about your storage space needs. Map data is not small. For example, a single map data file can easily take up 5–10MB of space.

Sweet emulation

Some mapping programs allow you to copy map data from a CD/DVD to your hard drive. This is useful because then you don't need to find misplaced data discs or swap between discs to view different maps. Unfortunately, some commercial mapping programs don't give you this option.

Skirt this problem with a special type of software: a *CD/DVD emulator*. This program allows

you to copy the contents of a CD/DVD onto your hard drive to create a virtual disc. This tricks a mapping (or other) program into thinking that you inserted its disc when the data actually already resides on your hard drive. Sweet. Do a Google search for *CD emulator* or *DVD emulator* to find information about various products. They tend to be reasonably priced — under $40.

If you plan to collect lots of map data, you'll definitely need a high-capacity hard drive for storage. At 10MB per data file, 100 files quickly can consume a gigabyte of disk space. Although you can get by using smaller hard drives, at a minimum, I'd suggest at least a 100GB drive.

If you decide to get serious about computer mapping, I recommend that you purchase a second internal or external hard drive to store your data exclusively. A second drive provides more performance and is easier to maintain and manage files. Plus, hard drives are pretty darn cheap these days.

CD and DVD drives

Just about every commercial software manufacturer uses CDs or DVDs to distribute their products. Digital map manufacturers are no different and extensively use CDs for map data. For example, I have a copy of National Geographic Back Roads Explorer in my library that came with a whopping 16 CDs that provides topographic map coverage of the entire United States. Fortunately, as DVD drives have become common in PCs, manufacturers are turning to DVDs for distributing large amounts of map data.

Having a DVD drive that can write *(burn)* DVDs or CD-ROMs is very useful if you plan to download large amounts of map data from the Internet. Because data files can be very large, archive the data on CDs or DVDs instead of cluttering your hard drive with infrequently used files.

DVD drives can read both CDs and DVDs. DVDs rock for storage because they can store a whole lot more data than a CD; compare 4.7GB on a DVD versus a relatively paltry 700MB on a CD.

Display Equipment

Obviously, you're going to want to see your maps, so you'll need a few pieces of display equipment, such as a graphics card and monitor. Again, these don't have to be anything fancy. (Are you noticing a theme here when it comes to mapping software hardware requirements?) Here are a few considerations for display hardware.

Make sure you have the latest drivers for your display hardware. This is especially important if you'll be rendering 3-D map images.

DVD soup

As this book goes to press, DVD standards have yet to be agreed on (by the manufacturers as well as the ever-important consumers). Check out this alphabet soup: There are DVD-ROM, DVD-R, DVD-RW, DVD+R, DVD+RW, and DVD-RAM formats. Some manufacturers are hedging their bets on the standards race by offering multiformat DVD drives. If you're in the market for a DVD drive, I'd certainly look for a multiformat drive until a single DVD standard emerges.

Graphics cards

Unlike computer video games, graphics card requirements for mapping programs are pretty minimal. All you need is a Super Video Graphics Array (SVGA) card, which has come standard on computers for years. If your mapping software supports 3-D terrain display, a card that has a graphics accelerator will draw map images faster. An accelerator isn't an absolute requirement because most commercial programs that support 3-D rendering take advantage of a graphics accelerator only if it is present. Check the specs of your current PC (or one you're considering purchasing) to see whether you have an accelerator.

Monitors

Just like the broad, general hardware requirements theme for mapping, bigger (that is, a bigger monitor) is better. Although most programs work fine on smaller screens, the larger the monitor, the more map area can appear onscreen. These days, the standard 17-inch and 19-inch monitors are more than adequate for digital mapping. However, if you're spending a lot of time using maps in front of a computer screen, consider a larger monitor, which is both easier on the eyes and can display much more data. (If you really get into maps, you can do like the pros and use two widescreen monitors, set side by side.)

In Windows, change the display size of a monitor via the Display Properties dialog box. To access this, right-click the desktop and select Properties from the contextual menu that appears. On the Settings tab, you can change the display area to different sizes, such as 800 x 600, 1024 x 768, or 1280 x 1024 pixels. Try some of these different settings to see which works best for your mapping program as well as your eyes.

Printers

At some point, you're probably going to want to print a digital map. (See Chapter 21 for some great tips on printing maps.) Expensive plotters and large format printers are important for a professional mapmaker, but any printer that can output the map in a readable fashion is fine for the average computer user. However, some printers are more suited for digital mapping than others.

✔ **Type:** Laser printers are best if you're making maps you'll be using outdoors — maps printed with inkjet printers have problems when they get wet; although special paper is available, which I discuss in Chapter 21. If you need to make maps outside your home or office, consider a small battery-powered inkjet printer.

✔ **Color printers:** Black-and-white printers are perfectly suitable for printing maps, but color output is usually easier to read and understand, especially when using topographic maps. A colored map produced on a cheap inkjet printer might be more useful than a crisp black-and-white map that came from an expensive laser printer. When cartographers make maps, they design them to be in either black and white or color. Important information can be lost when a map program translates a color map into the inherent shades of gray in black and white.

✔ **Resolution:** The higher the print resolution in dots per inch (dpi), the better the map will appear; especially for maps that show a lot of detail. Printers designed for printing digital photos work quite well in representing topographic and other detailed maps.

✔ **Print speed:** Some printers are faster than others, and a faster printer means you get to see and use your printed map quicker. Printers are rated in *pages per minute* (ppm), which is the number of pages that can be printed in a minute. When you're comparing page per minute ratings, be sure you look at the numbers for printing graphics instead of the numbers for printing text.

✔ **Cost per page:** If you're frequently printing maps, it makes good economic sense to use a printer with a reasonable *cost-per-page rating* (the estimated cost to print a single page, considering paper and ink). Cost-per-page rates vary considerably between printers and are usually mentioned in magazine and online reviews.

If you need to print a large map, check out your local blueprint or copy shop. If you can get the map into a TIFF or PDF format, they can print a super-sized version for you.

Communication Capabilities

If you want your GPS receiver to talk to your map software or if you plan to access the Internet to use Web map sites or to download map data, your PC needs to have some basic communication capabilities. Here are some communications ports and Internet connectivity considerations.

Communication ports

You can connect most older GPS receivers to your computer through the computer's serial port. (I discuss ports and connectivity thoroughly in Chapter 10.) If you want to download data from a GPS receiver to use with a digital map or upload maps and data to a GPS receiver, your computer needs a serial port and a special cable to connect the two devices.

Newer GPS receivers use a standard, faster USB connector to interface with a computer, which makes life much easier.

If your computer doesn't have a serial port (as is the case with a number of laptop computers) and you need one for your GPS unit, get a USB adapter. Read more about this in Chapter 10.

If your GPS receiver supports a serial or USB connection and using Secure Digital (SD) or MultiMediaCard (MMC) memory cards for storing data, use the memory card when you're exchanging data. It's both faster and easier to use when interfacing the GPS receiver to your computer. You will need a card reader connected to your computer to transfer data to and from the memory card. See Chapter 10 for more on memory cards and readers.

Internet connection

You need a modem and an Internet account if you want to

- Download freeware and shareware mapping programs and GPS utilities
- Download software updates
- Use Web-hosted mapping services
- Download data for creating maps

An Internet connection is a must for anyone interested in digital mapping.

If you plan to download digital maps and aerial or satellite photographs, you really should have a broadband Internet account (DSL or cable modem). Even when compressed, map files can be very large, and waiting for the data to arrive can be painfully slow on a dialup connection.

Chapter 10

Interfacing a GPS Receiver to a Computer

· ·

In This Chapter

▶ Transferring data between GPS receivers and computers

▶ Dealing with cables

▶ Understanding communications settings and protocols

▶ Using GPS receiver utility programs

▶ Upgrading GPS receiver firmware

· ·

*A*dmittedly, your GPS receiver is pretty darned useful even if you can't connect it to your computer. With your receiver, you can find your way around as well as your way back (without even knowing what a computer is). However, the ability to connect a GPS receiver to a computer can make your electronic navigator even more useful . . . heck, a *lot* more useful. Essentially, a connection between your GPS receiver and your computer allows them to talk and share information, such as maps and waypoints uploaded to your receiver as well as receiver information downloaded to your PC that documents your adventures.

In this chapter, I show you why this interconnectivity is a fantastic addition to your receiver's functionality. You discover the ins and outs of cable connections, the lowdown on great receiver utility programs, how to get your computer and GPS unit to speak the same language, and how to upgrade your receiver firmware.

About (Inter) Face: Connectivity Rules

Most handheld GPS receivers can interface with PCs these days (a few older models don't, and if you have one, I suggest it's time to invest in a new unit). Some of the very cool things you can do with a PC-compatible receiver include

✔ Back up and store GPS receiver waypoints, routes, and tracks on your computer. (Get the skinny on all these important critters in Chapter 4.)

✔ Download waypoints, routes, and tracks from your GPS receiver to your computer to use with computer mapping programs.

✔ Upload waypoints, routes, and tracks to your GPS receiver from other sources such as Internet sites, other GPS users, or mapping programs.

✔ Upload maps from your computer to your GPS receiver (if your receiver supports mapping). For more on selecting a handheld GPS with mapping capabilities, see Chapter 5.

✔ Provide GPS data to a moving map program on a laptop for real-time travel tracking such as the ones that I describe in Chapters 13 and 14.

✔ Update your GPS receiver's firmware.

This chapter is about interfacing GPS receivers (primarily handhelds) to PCs using cables and memory cards. There are GPS products that can communicate wirelessly with PCs. I discuss these in Chapter 7.

Anatomy of a Link: Understanding the Interface Process

Before I talk about how to interface a GPS receiver to a PC, you need to understand the types of data that can pass between the two devices:

✔ **GPS receiver to PC:** Saved waypoints, routes, tracks, and current location coordinates

✔ **PC to GPS receiver:** Maps (if the GPS receiver supports them), waypoints, routes, and tracks

You can interface a GPS receiver to a computer and transfer data in two ways:

✔ **Cable:** Most GPS receivers use a special cable, with one end that plugs into the receiver and the other that plugs into the serial or Universal Serial Bus (USB) port of your computer. (See the upcoming section "Untangling Cables.")

✔ **Memory card:** A growing number of GPS receiver models use Secure Digital (SD) or MultiMediaCard (MMC) memory cards to store maps and data. You transfer data between the GPS receiver and your computer with a card reader connected to the computer. (See the later section "Managing Memory.")

TIP

If you use a Bluetooth wireless GPS receiver, you don't need a cable or memory card reader to transfer data. These units are designed to be used exclusively with laptops and PDAs; I discuss these in Chapter 7.

Untangling Cables

If your GPS receiver uses a cable to connect to a computer and you want to interface the two, you need the right type of cable. Some GPS receivers come bundled with this cable, especially models with USB support; others don't.

If you don't have a cable for your GPS receiver, get one. Cables tend to vary in design between manufacturers and models, so be sure to get the right one for your GPS receiver. (See Figure 10-1 for examples of different types of cables.) You can purchase a cable directly from the manufacturer at its Web site or through a retailer. Expect cable prices to range from $20–$45.

Figure 10-1:
GPS
receiver
cables for
different
models.
USB cable
(top left),
USB serial
port adapter
(bottom left),
and various
serial cables.

Cables for older serial GPS receivers are called *PC interface cables.* They have a connector that attaches to the GPS receiver on one end (connectors vary between manufacturers and sometimes even models) and an RS-232, 9-pin connector on the other end that connects to your computer's serial port.

Newer GPS receiver models support USB. They typically feature a cable with a mini-USB connector at one end that plugs into the receiver, and a standard connector at the other end that plugs into a PC's USB port. If your GPS receiver supports both serial and USB interfaces, use a USB cable for much faster communications. USB GPS units also receive power from the cable when connected to a computer so you aren't running on the unit's batteries.

If you have a GPS receiver that's powered by or has its batteries recharged by a USB cable, you don't always need a computer nearby. There are wall socket and car cigarette lighter adapters you can plug a USB cable into that will provide juice.

If you bought a GPS receiver that didn't come with a serial cable (especially an older model) and you can't find a cable to buy, you're not necessarily completely out of luck. If you're handy with a soldering iron, most GPS receiver manufacturer Web sites describe the pin-out configurations of their cables so you can make your own. The tricky part can sometimes be finding the right connector for the GPS receiver because most connectors vary between manufacturers and models. A number of how-to sites on the Internet show you how to build your own cables and where to get the connectors. Do a Web search for *gps cable connector* and the brand of your GPS receiver to find different options.

Some serial cables are designed to power a GPS receiver from a cigarette lighter and to connect to a computer to send and receive data. These cables are especially useful if you're using your GPS receiver with a moving map program and a laptop. Just remember that you'll also need to buy a dual, car cigarette-lighter adapter so you can plug in both your GPS receiver and laptop into a single cigarette lighter.

If you have a Garmin receiver, check out the Pfranc company for its quality Garmin-compatible serial cables. Larry Berg started out making shareware Garmin cable plugs, and his business grew. He now stocks a line of reasonably priced cables for all Garmin models. Check out his Web site at www. pfranc.com.

Make sure the rubber covers that snap over your GPS receiver connectors are pressed into place and are in good shape. This is a vulnerable place on a receiver, and if water gets in an unsealed connector it can short circuit internal electronics. This usually means it's time for a new GPS receiver.

Pondering Protocols

A *protocol* is a way for two devices to talk with each other successfully. Think of a protocol as a language with a strict set of rules. When one device sends a message to another device, it expects a certain type of response back. This structured, back-and-forth takes place until one device sends a message that says the conversation is over.

Likewise, when you connect a GPS receiver to a computer, a certain protocol is used to transfer data between the two devices. You need to ensure that the same protocol has been selected for both devices. If two different protocols are used, it's like the GPS receiver speaking Russian to a computer that understands only Chinese.

The protocols typically used with GPS receivers are

- ✔ **NMEA:** The National Marine Electronics Association came up with the *NMEA 0183* standard, which is a protocol for transferring data between marine-related electronics, such as GPS receivers, autopilots, and chart plotters. Virtually all GPS receivers support the NMEA 0183 standard, which uses widely documented text messages. Typical NMEA data includes latitude, longitude, time, and satellite status.

  NMEA comes in several different versions, including 1.5, 2.1, 2.3, and 3.01. Make sure that this version number matches both the GPS receiver and the computer program that you're using.

- ✔ **Proprietary:** Some GPS manufacturers have proprietary protocols for communicating with a GPS receiver. These protocols send additional data that isn't included in the NMEA standard, for example, altitude, speed, and position error.

NMEA is the de facto standard for getting a GPS receiver to talk to a computer. However, if you have a choice between NMEA and a proprietary protocol (for example, the Garmin protocol used with Garmin GPS receivers), select the proprietary protocol because it can supply richer data to a program.

Understanding Serial Ports

Argg. You tracked down the right serial cable you need (if your GPS receiver comes with a USB cable, jump to the next section, "All About USB Ports"). You have the protocol business all figured out. You plug one end into your GPS receiver and the other into your computer. And nothing happens. Exactly. Nothing should happen because you need to be running some type of software on your computer that enables the two devices to talk to each other.

Before I discuss interface software, however, I have to lay some groundwork. The programs designed to communicate with GPS receivers have one thing in common: You need to specify certain communication parameters in both the program and the GPS receiver for the devices to exchange data successfully. If the settings aren't correct, you may as well try to communicate with someone a thousand miles away by sending smoke signals through a telephone line. Although setting the right communications parameters isn't that difficult, it can be a bit confusing. This section guides you through the process and gives you some tips on smoothing out some common problems that you may encounter, beginning with ports and their settings.

Setting up communications between a GPS receiver and a PC is a one-time process. After you get everything working, no worries about the next time.

COM ports

A *COM port* is a computer serial port that's used to connect a mouse, modem, or other device, like a GPS receiver. (*COM* stands for communication, and the ports are called *serial* ports because they receive data *serially,* one character at a time.) COM ports typically have a small oval, D-shaped connector (with nine pins) and are located on the back of your computer. (If you have other devices plugged into your serial ports, you'll need to unplug them so your GPS receiver cable has a port to plug in to. You can plug the other devices back in when you're through transferring data with your GPS receiver.)

Your PC might have one or two physical serial ports that you can plug devices into, but Windows allows you to assign a COM port number to each device. These numbers usually range from 1 to 4 but might go as high as 256 if a USB adapter is used, (which I talk about in the upcoming section, "All About USB ports"). In most cases, you won't need to use Windows to reassign any of the port numbers. Just know that you have numbered COM ports and that you need to assign one to your GPS receiver, which I talk about next.

To get more detailed information about COM ports and Windows, including how to change settings with Windows Device Manager, head to the Microsoft support Web site at `http://support.microsoft.com` and search the Knowledge Base for *com port.* Or pick up a copy of *PCs For Dummies* (Wiley Publishing, Inc.) by Dan Gookin, which has an excellent chapter that clearly explains everything you need to know about the subject.

The program you're using to interface with the GPS receiver is where you need to specify which COM port number the receiver is connected to. Programs typically use a drop-down list that shows all the COM ports; just select one from there. (Some programs have an autoselect feature that tries to establish communications on all available COM ports until the port with the GPS receiver is found.)

COM ports have properties that establish how the communication between the two devices will occur. Generally, both of the devices need to have the same settings. The COM port properties include

- ✔ **Baud rate:** *Baud* is the speed at which the port communicates with other devices. This number is in bits per second (bps): the bigger the number, the faster the speed.

- ✔ **Data bits:** This is the number of data bits that are transferred for each character, typically 7 or 8.

- ✔ **Parity:** This is a form of error checking that ensures the integrity of transferred data.

- ✔ **Stop bits:** This is how many bits follow a character and mark the end of a data transmission.

- ✔ **Flow control:** Sometimes called *handshaking,* this is a way for one device to stop another device from sending data until it's ready to receive the data.

Although you can set COM properties in the Windows Device Manager, I recommend making changes in the program that you're using to interface with the GPS receiver. You'll usually see an Options or Configuration menu in the program that displays a dialog box where you can set these values. (See Figure 10-2 for an example of a communications setting dialog box in a mapping program.)

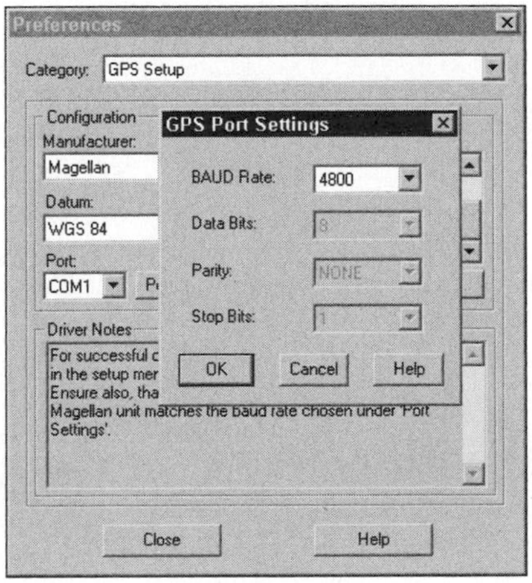

Figure 10-2:
Set COM
properties
to match
your
interface
program.

Unless the program specifies otherwise, here are the typical COM port settings to use when interfacing with a GPS receiver:

- **Baud rate:** 4,800 and 9,600 baud are almost certainly guaranteed to work with all GPS receivers. You can increase the speed to a higher rate on some types of GPS receivers. The higher the speed, the faster the data transfer. I recommend experimenting until you find the fastest, most-reliable baud rate, and then using that setting.

- **Data bits:** 8

- **Parity:** None

- **Stop bits:** 1

- **Flow control:** None

Some GPS receivers allow you to set the baud rate in the system setup page of the receiver, but other models don't give you any control of the speed. Check your GPS receiver user guide for more information about your model.

All About USB Ports

Consumer GPS receivers first started appearing in the mid-1990s. At that time, personal computers exclusively used serial ports to interface with other devices, making it easy for hardware manufacturers to design their products to communicate through a serial port. GPS was initially popular with sailors because they could connect a GPS receiver to an autopilot or chart plotter and navigate a vessel based on GPS data. With the right cable, you could also connect your GPS receiver to a computer and download and upload data.

Fast forward to the present. Many newer GPS receivers come with USB cables to connect to a PC. The computer industry's goal with USB was to make it incredibly easy to interface hardware devices to a PC. None of that setting baud rates, parity, and stop bits stuff. Just plug and play. So how do GPS receivers fit into the scheme?

Virtual serial ports

Most USB-compatible, handheld GPS receivers come with special driver software that creates a virtual serial port associated with the GPS unit. You install and configure the driver software (which comes with the receiver on a CD-ROM or DVD). When you plug the GPS unit into your computer with a USB cable, the software emulates a serial port connection. This allows third-party mapping software that needs to communicate with a GPS unit through a serial port to do so successfully even though there's a USB connection (which uses a different protocol). Check your GPS unit's user manual for more information.

Always make sure you have the latest USB driver, the latest version of the mapping software (that supports your receiver), and the latest version of the GPS receiver firmware.

Serial ports are pretty much going the way of the dinosaur, replaced by the easier-to-use and faster USB ports. In fact, some laptops no longer have serial ports.

Some USB GPS receivers use the slower (12 Mbps) 1.1 USB standard. This contrasts with the faster (480 Mbps) 2.0 standard found in most USB printers and hard drives.

USB serial port adapters

What do you do if you have a GPS unit that can only connect to a computer with a serial port and your computer doesn't have one? The solution is to use a USB serial port adapter. The adapter plugs into your computer's USB port and has a standard 9-pin connector that you can connect serial port devices to. After you install a driver for the adapter (which comes with the product on a disk or CD), Windows recognizes the adapter as a serial port. Just connect a PC interface cable to your GPS receiver and the adapter, and you're all set to send and receive data between the two devices. (See a USB adapter in Figure 10-1.)

Note this one little "gotcha" to mention regarding USB adapters: Windows might squawk that you need a driver when you plug in the adaptor, but you know you've already installed one. The fix: In the dialog box that's prompting you for the driver location, tell Windows to look in the `C:\Windows\Drivers` directory. (This path is for Windows XP; the location varies in older versions of Windows: for example, `C:\WINNT\system32\drivers` in Windows 2000.) Depending on what USB devices are running at the time, Windows XP might assign the adapter to a different COM port from the last time it was used and incorrectly assume it's a new device that needs a new driver. This above-mentioned directory is where previously installed drivers are stored, saving you from having to find the original driver distribution disk.

All GPS receiver manufacturers sell their own branded serial-to-USB adapters, albeit at a premium price. If you're on a tight budget, most any third-party adapter that you buy from a computer retailer will work. These adapters tend to be cheaper than the GPS manufacturer brand-name models. I've found the Keyspan USA-19HS adapter to work quite well with a number of different GPS units and map programs.

Managing Memory

If you own a GPS receiver that uses a memory card, congratulations! I personally like the versatility these receivers offer (such as stuffing a bunch of maps onto a single, large-capacity memory card). This section talks about how to get the most out of your memory card GPS receiver. If your GPS receiver uses only a cable to connect to a computer, you can skip this section. Better yet, read along to see how memory cards work for transferring data. (Figure 10-3 shows a GPS receiver with a memory card.)

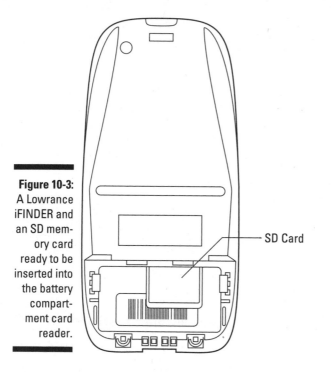

Figure 10-3: A Lowrance iFINDER and an SD memory card ready to be inserted into the battery compartment card reader.

— SD Card

If your GPS receiver supports using a memory card, you have some significant advantages when exchanging data with a computer, including

✔ **Upload speed:** Uploading maps from your computer to a GPS receiver is considerably faster with a memory card than via a serial port cable or even with a 1.1 USB connection. Because GPS receivers communicate at a fairly low baud rate, transferring 10–20MB of map data can take a long time (up to hours depending on how the serial port is configured).

With roughly an 80 Mbps top-end transfer rate, writing map data directly to a microSD card is much faster than writing the data to your GPS unit through a 12 Mbps USB 1.1 connection. If you have a GPS receiver that uses microSD

cards, you can speed up transfers even more by purchasing a microSD-to-SD adapter and a USB 2.0 SD card reader. This allows you to take advantage of USB 2.0's faster 480 Mbps maximum transfer speed.

- **Affordable and practical:** Memory cards have gotten inexpensive over the years. You can load frequently used maps on several cards and not bother with repeatedly uploading data from map program CDs. Expect to pay around $10 per gigabyte, or even less, for standard memory cards.

- **More storage:** Memory cards give you considerably more storage than GPS receivers with fixed amounts of internal memory. You can currently purchase memory cards that provide between 512MB–8GB of data storage (with that maximum number seemingly increasing all the time). Handheld GPS receivers that don't support memory cards might have only 1MB to under 200MB of internal storage.

Some GPS units that support memory cards might have an upper limit on the maximum size of the card. That means that just because you have an 8GB SD card, it doesn't mean you can effectively use all that space. Because of hardware limitations, your GPS receiver might be able to support only a 1 or 2GB card. Check your user manual before purchasing a large capacity memory card. Also, check that newer high-speed memory cards are compatible with your receiver.

- **Versatility:** You can use the memory card in your PDA, digital camera, and cellphone (if all the devices support the same type of card).

- **Minimal settings:** When using a memory card, you don't need to worry about COM ports, baud rates, and protocols when transferring data. (However, if you're using your GPS receiver with a laptop and cable connected to a moving map program, you still may have to contend with getting all the settings just right.)

Here are a few drawbacks to GPS receivers that support memory cards:

- **Added cost:** They add a bit more cost to the receiver price because of the built-in card reader and associated technology.

- **Reader:** You need a memory card reader connected to your computer to transfer data back and forth between the GPS receiver. However, some computers (notably laptops) have built-in card readers.

If your GPS receiver didn't come with a card reader, unless the manufacturer states otherwise, any third-party reader will work. These devices are inexpensive and easy to use. Just plug the reader into a USB port, and Windows treats the memory card like a hard drive or floppy disk. You can then copy data back and forth between your hard drive and the memory card. Card readers are inexpensive, and you can purchase a basic model for under $20.

✔ **Removal:** Memory cards can be a little tedious to swap because you need to remove the GPS receiver batteries to access the card slot. (The battery compartment is the only easily accessible, waterproof location on a receiver.)

Just like digitals cameras, GPS receivers that support memory cards usually come with a card that has a relatively small amount of storage space. In fact, if they both use the same type of memory card, you can swap a card between your digital camera and GPS receiver. You'll probably want to upgrade to a larger capacity card. Blank memory cards with the GPS receiver manufacturer's brand name tend to be more expensive than standard MMC and SD cards. I don't find any difference between the two, and you can save money with third-party memory cards in your GPS receiver.

Transferring GPS Data

After you have a cable and then get the ports, baud rates, and protocols all figured out (or have a memory card and card reader), the next step is getting the data transferred between the GPS receiver and your computer. This is where software comes in, and you generally have three options:

✔ **GPS manufacturer software:** The commercial mapping programs that GPS receiver manufacturers sell can all upload and download waypoints, routes, and tracks as well as upload maps to certain models of their GPS receivers. (Using GPS manufacturer software packages and their features is presented in Chapter 11.)

GPS receivers that display maps typically work only with proprietary maps provided by the manufacturer. You can't upload maps from third-party mapping programs into your receiver. Sometimes GPS novices believe that they can load maps from other companies or sources directly into their GPS receiver. In most cases, you can't. This may be starting to change though. DeLorme's handheld Earthmate PN-20 GPS unit supports user-created maps (including aerial and satellite photos) using the company's XMap software. Certain Magellan models also support maps uploaded from the National Geographic Topo! Map products.

✔ **Third-party mapping software:** Most third-party, commercial map programs can also upload and download waypoints, routes, and tracks. This is an essential feature so you can overlay GPS data on a digital map appearing on the PC to see where you've been. You can also plan a trip with the mapping software and then upload waypoints and routes from your computer to a GPS receiver.

✔ **GPS utilities:** Software programs used to interface GPS receivers to computers are utility programs designed specifically to download and upload waypoints, routes, tracks, and other information. These programs are usually freeware or shareware and have a number of useful features.

GPS receiver interface programs tend to work the same way, although they have different menus, dialog boxes, commands, and appearances.

When you transfer data between your GPS receiver and PC, you'll either be

- Sending current location coordinates to use with a real-time mapping program you have running on your laptop (or PDA). As you move, your location appears on the map.

- Downloading or uploading waypoints, routes, tracks, or map data.

The process to complete both of these tasks is the same. (If you have a GPS receiver with a memory card, you don't need a cable to download or upload waypoints, routes, tracks, or map data because you'll be using the memory card and a card reader to do this. However, you'll still need a cable to send your current location to a real-time mapping program.) Take a look at the general steps involved in transferring data with a cable:

1. **Connect the interface cable to your computer and to your GPS receiver.**

2. **Turn on your GPS receiver.**

   The GPS receiver doesn't need to have a satellite fix to transfer data unless you're using the receiver with a mapping program that's plotting your current location.

3. **Run the interface program.**

4. **Ensure that the protocols and settings on both the GPS receiver and the computer are the same if you're using a serial connection.**

At this point, you can

- Select the type of data (waypoints, routes, or tracks) and upload it to the GPS receiver from your computer or download it from the GPS receiver to your computer.

- Upload a map to the GPS receiver that was created with a GPS manufacturer's mapping program.

- Have the GPS receiver start providing location data to the interface program for real-time mapping.

Check the program's user manual or online help for specific instructions on transferring data between your GPS receiver and a PC. If the program can't connect to the GPS receiver, read the upcoming "Troubleshooting Connection Problems" section.

If you're transferring data to and from a memory card, refer to the GPS receiver's user manual or support Web site. With some receivers, you can simply drag data between the memory card and the hard drive via Windows Explorer. In other cases, you might need to use a utility program, such as G7ToWin or EasyGPS. (See the sidebar, "G7ToWin and other utilities.")

G7ToWin and other utilities

G7ToWin is the Swiss Army knife of the GPS world. This free utility program works with many brands of popular GPS receivers. With this Windows program, you can

- Upload waypoints, routes, and tracks to your GPS receiver.

- Download waypoints, routes, and tracks to your computer.

- Edit waypoint, route, and track data in a spreadsheet-style window.

- Create waypoints, routes, and tracks.

- Save waypoints, routes, and tracks in different file types. (Remember that GPS receiver and software manufacturers all use a number of different file formats for waypoints, routes, and tracks.)

- Save the image that appears onscreen of many Garmin GPS receivers.

G7ToWin is a must for the serious GPS receiver user. Its author, Ron Henderson, continues to add new features to the program. DOS and Pocket PC versions of the utility are also available. You can download G7ToWin at www.gpsinformation.org/ronh.

Other GPS utilities worth consideration include

- **EasyGPS:** A popular, free utility for creating, editing, and transferring waypoints and routes, available at www.easygps.com.

- **GPS Utility:** A popular freeware/shareware program that lets you manipulate and map waypoints, routes, and tracks and works with Geographic Information System (GIS) data. It's available at www.gpsu.co.uk.

- **GPS TrackMaker:** A free utility that creates, edits, and deletes waypoints, routes, and tracks. It supports mapping features. You can download the program at www.gpstm.com.

- **GPS Babel:** The multi-platform wonder tool. It slices, it dices, it converts. A must have if you're a techie. Open source and free at www.gpsbabel.org.

- **XImage:** If for some reason you can't do a Garmin product screen capture with G7ToWin, try the utility that comes straight from the manufacturer: www.garmin.com or use Google.

- **GPSy:** For users who want to interface their receiver with a Mac, see www.gpsy.com.

Troubleshooting Connection Problems

If you follow the instructions that come with the mapping and utility software, getting the GPS receiver and computer to talk to each other is usually painless.

If you do run into problems, follow this set of steps, in this order, to help you identify a possible culprit for your connection troubles:

1. **Always make sure the cable is securely plugged in to both the GPS receiver and the computer.** While you're at it, check that the GPS receiver is turned on.

2. **Make sure that the serial port baud rate and the protocol are the same in both the GPS receiver and the interface program.** Double-check this again if you can't establish a connection.

   Even if the baud rates match, they may be set too high — thus causing communication errors. When in doubt, lower the baud rate. You can either step-down a rate at a time or go directly to 4,800 or 9,600 baud. Although this is slow, these rates shouldn't generate errors.

3. **In the interface program, make sure that the correct COM port is specified.** If you can't get a connection, try different COM port numbers until you find one that works.

4. **Do you have the latest USB driver, the most recent version of the mapping software, and the current release of the GPS receiver firmware?**

5. **Always check the program's user manual, online help, or support section of the vendor's Web site for specific information on interfacing with a GPS receiver.**

If you can't get your GPS receiver to talk to your computer and you happen to have a PDA, turn off the PDA synchronization program first. PDA synchronization software that's running in the background is a frequent culprit in causing GPS receiver interface problems.

Uploading Firmware Revisions to Your GPS Receiver

Just like software vendors, GPS manufacturers find bugs and add enhancements to their products. New versions of a GPS receiver's operating system can be upgraded through the receiver's *firmware* (the updateable, read-only software that's embedded in a hardware device).

Check that your GPS receiver's firmware is current every few months or so, especially if your receiver is a newly released model. GPS manufacturers offer free downloads of firmware upgrades on their Web sites, and these bug-fixes or new features can definitely make your GPS receiver perform better.

Here's how to upgrade your firmware:

1. **Check the current version of your GPS receiver firmware.**

   Sometimes this displays when the GPS receiver is turned on, or it might be on an information page. Consult your user's guide or the manufacturer's Web site for specific instructions on how to get this information for your model.

2. **Visit the manufacturer's Web site and go to the software updates section.**

   Here are the URLs of the major handheld GPS manufacturers:

   - **DeLorme:** www.delorme.com
   - **Garmin:** www.garmin.com
   - **Lowrance:** www.lowrance.com
   - **Magellan:** www.magellangps.com

   Some manufacturers have e-mail notification when new firmware is released as well as programs and Web pages that check your firmware to see whether it's up-to-date. If it isn't, download the correct installer file.

3. **Find your GPS receiver model and check its manufacturer's Web site for the latest firmware version.**

   If your firmware is older than the current version on the Web site, follow the online instructions to download the firmware installer. Usually, the higher the version number, the more recent the firmware version.

   Make sure that the firmware installer you download is for your GPS receiver model. If you upload firmware designed for a different model, plan on the GPS receiver not working until you load the correct firmware.

4. **Follow the installation instructions that come with the downloaded file.**

Usually firmware installation files come in two forms:

- ✔ **A standalone program** that runs on your computer, connects to the GPS receiver, and sends the upgraded firmware to the receiver. You need to have a cable attached to both the computer and the GPS receiver.

- ✔ **A special file** that you copy to a memory card. When the GPS receiver starts, it searches the card to see whether a firmware upgrade is present. If it is, the receiver uploads the upgrade. After the upgrade is successful, you can erase the firmware upgrade file from the memory card.

Upgrading a GPS receiver's firmware is pretty easy; not too much can go wrong. About the only thing that can get you in trouble is if the GPS receiver's batteries die midway through a firmware upload. A firmware upgrade usually only takes a few minutes to complete, but make sure that your batteries aren't running on empty or that you're plugged into an external power source before you start. If something does go wrong during a firmware upgrade, you'll likely need to perform a system reset. This usually involves pressing some button combination that varies between manufacturer and model — use Google or consult some of the online resources I mention in Chapter 20 for more information.

Some firmware update software works only on COM ports 1 through 4. If you're using a USB adapter (which is usually set to COM port 5 or higher) and are having problems connecting to the GPS receiver, try reassigning the existing COM ports to numbers higher than the USB adapter's port; then set the adapter's port number to 1. Refer to online Windows help (choose Start⇨Help) and perform a search for *device manager* to get more information on changing device settings.

Chapter 11

Using GPS Manufacturer Software

*I*n the not so distant old days, GPS receivers just showed latitude and longitude coordinates and displayed very simple maps with virtually no details. Times have changed, and GPS units are now capable of providing all sorts of information as well as displaying a wide variety of maps and charts.

GPS receiver manufacturers offer many different types of maps — street maps, topographic maps, fishing maps, nautical charts, and more. These commercial map products often come with software to plan trips, exchange data with your PC, and install the maps on your GPS receiver. In addition to maps, many manufacturers also offer free utility programs that interface with their GPS receivers and perform various useful functions.

In this chapter, I discuss map and utility software for handheld GPS receivers produced by Garmin, Magellan, Lowrance, and DeLorme. I also show you the general features that all GPS map programs share as well as what kind of maps are available for the different GPS receiver brands.

Because manufacturers don't offer as many choices for automotive GPS maps and software, I don't cover them in this chapter. If you have an automotive GPS receiver, check the manufacturer Web site for compatible maps and software.

Understanding GPS Map Software

When you purchase GPS map software, you're buying map data and a PC program designed to interface with your GPS unit. (A GPS receiver's ability to display maps depends on its firmware and hardware components, such as the processor and display.)

If you can, avoid paying list price for GPS map software. You can usually find map software at discounted prices from reputable online retailers.

For more information and in-depth reviews of GPS receiver manufacturer map products, including screenshots, visit `http://gpsinformation.net`.

Some GPS map software rules of thumb

It's important to be aware of a few rules of thumb that apply to all GPS units and map software, regardless of the manufacturer.

- **Map compatibility varies.** Your GPS unit model dictates what maps you use. Some handheld GPS receivers don't support uploadable maps. Certain models might be able to use the autorouting feature found in street map software, although others may not. You might be able to load a topographic map to a manufacturer's handheld receiver but not load the map to the same manufacturer's automotive GPS unit. My point is, before you purchase a map product make sure it will work with your GPS receiver. Manufacturer Web sites are a good place to check.

- **No mixing and matching.** Generally, you can only use map products sold by the manufacturer of your GPS receiver. That means you can't use a Magellan map on a Garmin GPS unit, or vice versa. Also, you can't take maps from your favorite PC map program and move them onto your GPS receiver. (I discuss a few exceptions to this rule shortly.)

- **How many maps you can store depends on available space.** The amount of internal memory or the size of the memory card (on certain models) determines how many maps can be stored on your GPS receiver. With a large memory card, you can easily store maps covering the entire United States.

- **Don't expect the level of detail in a GPS map to match paper maps.** In order to maximize memory space, most GPS receivers use *vector* maps (created with lines and shapes) instead of *raster* maps (scanned paper maps). That means a map on your GPS receiver won't appear as an identical version of your favorite paper map. (DeLorme is an exception to this rule, offering scanned USGS topo maps, NOAA nautical charts, and satellite and aerial photos that you can upload to their handheld receiver.)

Even if your GPS receiver doesn't support maps, you can still use a manufacturer or third-party map program with your receiver to download and upload waypoints, routes, and tracks, as well as access maps on your PC.

Common GPS map software features

Here are some features that all GPS receiver map software has in common.

- ✔ **Upload maps to GPS receivers:** The main job of a mapping program on your computer is to upload maps to your GPS receiver. The maps generally appear the same on your PC and your GPS receiver screen, although the GPS receiver displays smaller portions of the map than you can view on your computer screen. (And the map obviously won't appear in color if your receiver has a monochrome display.) You use the mapping software to select the portions of the maps that you want to upload to your GPS receiver. (To learn how to upload to your receiver, see the following section.)

- ✔ **Print from and use maps on your PC:** In addition to uploading maps to a GPS receiver, you can use the map software on your PC to view and print maps, measure distances, and plan trips.

- ✔ **Download waypoints, routes, and tracks from GPS receivers:** With GPS mapping software, you can download information that you've recorded with your GPS receiver, such as waypoints, routes, and tracks. You can store this data on your PC's hard drive or display it as an overlay on the maps displayed on your PC. (Read how to create and use waypoints, routes, and tracks in Chapter 4.)

- ✔ **Upload waypoints, routes, and tracks to GPS receivers:** In addition to downloading GPS data, you can also upload waypoints, routes, and tracks from your PC to a GPS receiver. For example, you can plot several waypoint locations on the PC map and then transfer them to your GPS receiver.

- ✔ **View POIs:** Some software packages have Points of Interest (POIs), such as restaurants, gas stations, and geographic features shown on the maps that you view with your PC or GPS receiver.

Getting GPS maps

GPS map software and data are distributed in a variety of ways, including

- ✔ **CDs and DVDs:** This is the most common way to purchase GPS receiver maps. You buy a CD or DVD with the map software and data, install it on your PC, and then transfer selected maps to your GPS unit.

- ✔ **Memory cards:** Many manufacturers give you the option of purchasing maps that are preinstalled on memory cards. This saves you the hassle of transferring maps from your PC to your GPS unit. No PC software comes with these cards; you simply plug them into your compatible GPS unit, and *voilà!,* instant map.

Some manufacturers don't allow you to use map data sold on memory cards with mapping software on your PC. This contrasts with maps purchased on CDs and DVDs, which can be used on a PC and uploaded to a GPS unit. Check manufacturer information before you buy.

Another downside to preinstalled maps on memory cards is if you lose or damage the card, you're out of luck and will need to purchase a new card. If you have map data on a CD or DVD, you can create a new map and save it to a memory card.

✔ **Downloads:** Some manufacturers allow you to purchase and download specific topographic maps and charts from Web sites. This makes sense if you only plan to use your GPS receiver locally. Why buy an entire region or the entire United States if you're just tromping around in nearby hills? (Street maps are an exception in that you need to purchase a full region or country.) You copy the downloaded map to a memory card or use a program to transfer it to your GPS unit.

✔ **Non-manufacturer sources:** Just moments ago, I said you could only use manufacturer-supplied maps on your GPS unit. That rule used to be cast in stone; however, times are changing, and you should be aware of a few exceptions.

• A growing collection of free Garmin-compatible maps isn't produced by Garmin (see the upcoming "Do-it-yourself maps" sidebar).

• National Geographic TOPO! PC software maps can load onto certain Magellan GPS receivers.

• DeLorme's XMap software allows you to load just about any kind of map you can think of onto the company's handheld receiver (see the section on DeLorme at the end of this chapter).

Faster processors and high-resolution screens allow some models of GPS receivers to display raster maps (such as scanned paper maps). That may seem pretty cool, but there are trade-offs. You don't have the almost infinite level of zooming you have with vector maps; raster maps take up considerably more storage space; screen redraws are slower; and faster chips often mean shorter battery life.

Loading maps to GPS receivers

All GPS manufacturer map programs get map data to your GPS receiver in similar ways. Although the user interfaces are different and the commands vary, you load a map onto a GPS receiver by using the same basic process.

1. **Run the map program and zoom in on the area that you want to upload to your GPS receiver.**

2. **Choose which parts of the map you want to upload to the GPS receiver.**

   This usually means selecting one or more rectangular areas on the map. Depending on the program, you either draw a rectangle that defines the areas that you want to upload, or you select grid squares that appear on the map that correspond to the areas you want to upload.

3. **Select whether to save the map to your GPS unit or write the data to a memory card.**

   If you're writing map data to a memory card, make sure you have a card inserted in the card reader/writer (some programs allow you to save the map to your hard drive). If you're transferring the data directly to your GPS receiver, ensure it's connected to the PC with a USB or serial cable and the communication settings are correct. (I talk about connecting GPS units to PCs in Chapter 10.) Even if you're writing the map data to a memory card, with some programs you still might need to have your GPS unit connected to your PC so the program can read the receiver's serial number.

   It's a lot faster to transfer map data directly to a memory card reader/writer than moving the data to your GPS unit via a USB connection.

The program extracts the information that it needs from the map data and builds a custom map of the area that you select. When this process finishes, the program either starts uploading the map to the GPS receiver or saves the map to a memory card or your hard drive — after it's on the drive, you can later copy it to a memory card.

How long building a map takes depends on the size of the area that you select, how much map detail you want to include, and how fast your PC is. This can range from a minute or less for small areas (such as a metropolitan area) to 15 minutes or more for a large map (such as one that includes many different states).

How much time it takes to upload a map to a GPS receiver also depends on the size of the area you select and how the receiver stores maps. If you're uploading a large map from a PC via a serial cable, it can take hours to transfer the map from your PC. GPS receivers that support Universal Serial Bus (USB) communications are much faster. For GPS receivers that use memory cards, after the map has been created and saved to the card, it's just a matter of inserting the card into the receiver.

After you purchase GPS map software, be sure to check the manufacturer's Web site every now and then to see whether updated releases of the PC software are available. You may be able to download upgraded versions of the program with bug fixes and enhanced features. Keep in mind that when you download the program, updated map data doesn't typically come with it.

Some GPS receiver manufacturers use different methods for stemming software and map piracy. Both Garmin and Magellan use *unlock codes* on some of their map products that require you to visit a Web page and get a code to activate the map data. Some map products (notably nautical charts) have multiple regions stored on a CD-ROM or DVD, and you need to purchase an unlock code for each region you want to access. In addition, programs commonly link the serial number of a GPS receiver to a map, meaning that the map will work only with the GPS receiver that the map was originally uploaded to. Also, some programs might limit the number of GPS units or memory cards you can write map data to.

Exploring Garmin Software

Garmin (www.garmin.com) is the largest consumer GPS manufacturer and produces many handheld, automotive, and specialty navigation products. The company also has an extensive collection of commercial map products as well as a variety of free utility programs you can download.

The Garmin Web site has a large amount of information and sometimes it can be a challenge finding what you're looking for (especially with the free utility programs I mention). To speed things up, use Google. For example, if you want to download a copy of xImage, a nifty little program for doing GPS unit screen shots, search Google for *garmin ximage* and then click the Garmin Web site link.

MapSource

MapSource is Garmin's main program for uploading data and maps to their GPS receivers as well as exchanging waypoints, tracks, and routes between a PC and GPS unit. The software comes packaged with most of the company's map products that I discuss. A screen shot of the interface is shown in Figure 11-1.

When it comes to software, GPS manufacturers have long neglected the Mac community, favoring Windows users instead. However, Garmin is making efforts to change that with the release of Bobcat, a program that offers MapSource functionality for the Mac. Garmin also has several other GPS utility programs for Macs available; see www8.garmin.com/macosx.

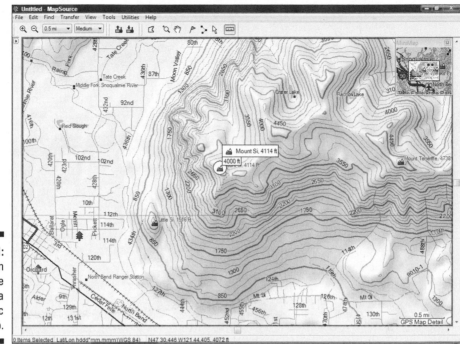

Figure 11-1:
Garmin
MapSource
displaying a
topographic
map.

Do-it-yourself maps

For the most part, GPS manufacturers have a lock on the market when it comes to maps that can be uploaded to their receivers. GPS receiver owners must use proprietary maps distributed and sold by the manufacturers.

However, a growing group of technically adept GPS and map enthusiasts has found ways around this map monopoly. These developers are turning out free maps you can upload to your Garmin receiver including maps for many locations where commercial maps aren't available and topographic maps that in some cases have even more detail than those sold by Garmin. These maps aren't limited to the outdoors. There's also work going on to make available free street maps (including autorouting capabilities) based on the Open Street Map project (www.openstreet map.org).

With programs such as cGPSmapper (www. cgpsmapper.com), Moagu (http:// moagu.com), and Mapwel (www.mapwel. biz), anyone can create a custom map to load on compatible, Garmin GPS units. (It's a bit of an involved process, but tutorials are available.)

If you're not into making your own maps, you can benefit from those that do. A large directory of free, user-created Garmin-compatible maps for locations all over the world can be found at http://garminmapsearch. com or www.gpsfiledepot.com for U.S.-only maps.

Garmin map products

Garmin offers a number of products that use MapSource to access different types of map data. Sold separately, there are packages for street, marine, and topographic maps.

- ✔ **City Navigator:** City Navigator contains detailed street maps and points of interest and is designed to work with Garmin GPS receivers that support *autorouting* (turn-by-turn street directions). Versions of the product are available for North America, Australia, Europe, the Middle East, New Zealand, South Africa, and Thailand. NT versions, which feature more highly compressed maps, only compatible with certain GPS units are available for North America, Europe, the Middle East, Brazil, Singapore/Malaysia, Australia and Mexico.

- ✔ **MetroGuide:** MetroGuide is similar to City Navigator but is designed for Garmin GPS receivers that don't support autorouting. Versions of this product are available with street map data for the United States, Canada, Australia, and Europe.

- ✔ **WorldMap:** This software provides basic international maps that expand the default basemap coverage that comes with your GPS receiver.

  GPS receivers sold in different parts of the world typically have different basemaps. For example, a GPS unit sold in Europe would have more basemap detail for Germany compared to a model sold in the United States.

- ✔ **BlueChart:** BlueChart products are Garmin's nautical charts and are available in regional versions that cover the Americas as well as the Atlantic and Pacific Oceans. Bathymetric charts, which show depth contours, are also available.

- ✔ **BlueChart g2:** g2 nautical charts show more detail and information. They're more expensive than standard BlueChart products and are only available on preprogrammed memory cards.

- ✔ **Inland Lakes:** Maps of popular fishing areas in North America with depth contours, shoreline details, boat ramp information, and fishing tips. The maps are sold for multi-state regions. Also, a version called Vision, containing 3-D bathymetric charts and satellite imagery, is available on preprogrammed memory cards.

- ✔ **Minnesota LakeMaster:** Designed specifically for GPS receiver owners who fish in Minnesota, this map product has maps of popular lakes and includes 3-foot contours with underwater structure detail, islands, reefs, points, bays, access points, and marinas.

- ✔ **MapSource U.S. TOPO:** The TOPO product contains 1:100,000 scale maps of the United States and shows terrain contours, elevation, trails, roads, and summits. It's designed for outdoor recreational use.

Garmin also offers topographic maps of Canada and Great Britain. Additionally, third-party partners offer many other international maps. See `http://www8.garmin.com/cartography/mpc` for details.

✔ **MapSource U.S. TOPO 24K:** 24K doesn't refer to gold but rather to 1:24,000 scale maps, which have significantly more detail than the maps found in MapSource TOPO. The maps provide detailed coverage of U.S. national parks and surrounding national forest lands. Points of interest and park amenities are also included. Three versions of the product are available: National Parks, East, National Parks, West and National Parks, Central.

Garmin also sells a product called Trip and Waypoint Manager that downloads, uploads, and manages waypoints, routes, and tracks. If you have a MapSource product, these features are built-in, so you don't need Trip and Waypoint Manager. Also, a number of free products, such as EasyGPS, are available on the Internet that offer the same functionality.

Garmin utility programs

In addition to map products, Garmin also offers a number of free software utilities for their GPS receivers.

Communicator

This free Windows and Mac Web browser plugin allows you to transfer data from compatible Web sites directly to USB Garmin receivers. By adding some code to a Web page, sites can exchange waypoints, tracks, routes, and other information with connected GPS units.

MapSource on the Web

If you use a Garmin GPS receiver and are interested in seeing what maps from the different map products look like — or to check out the amount of detail and coverage for certain areas — visit `www.garmin.com/cartography` and check out the MapSource Map Viewer.

Choose any of Garmin's map products from a drop-down list, and the selected map is shown in your Web browser. After the map displays, you can move around the map and zoom in and out. What you see on your PC monitor is what you can generally expect to see on your GPS receiver if you upload that particular map product. Remember that your GPS receiver screen is smaller; and, if it doesn't support color, the map will display in monochrome.

Even if you don't use a Garmin GPS receiver, this Web site is useful for getting a better idea of the types of maps that are available for GPS receivers, their general appearance, and what information they show.

nRoute

nRoute is a free Windows street navigation program designed to run on laptops connected to a Garmin GPS product. The software accesses MapSource data you already have on your laptop (such as from the MetroGuide or CityNavigator products) and provides directions and a moving map as you travel.

POI Loader

A growing number of third parties are providing POIs (Points of Interest) for specialized use, such as listing locations of red light and speed cameras, chain restaurants and stores, and historic places. The free POI Loader program uploads POIs in several different formats compatible to Garmin GPS units. You can even create your own POIs from a spreadsheet file.

Most Garmin GPS receivers have a maximum limit of 500 to 1,000 waypoints. However, the number of POIs is limited only by the space available on a memory card, so you can potentially have tens of thousands of POIs for navigation and information.

Check out GPS-POI-US (`www.gps-poi-us.com`), POI Factory (`www.poi-factory.com`), and POInUSA (`www.poinusa.com`) for lists of various POIs.

Spanner

Spanner is a free Windows program that reads current GPS coordinates from Garmin's proprietary data transfer protocol and translates the coordinates into NMEA 0183 format. This utility was originally designed for Garmin's USB mouse GPS product, so the receiver could interface with other navigation programs (such as Microsoft Streets & Trips, DeLorme Street Atlas, and OziExplorer) that use real-time NMEA data to display moving maps. Spanner also works with a number of other Garmin receivers, and allows you to use your handheld GPS unit with a laptop and navigation software.

Spanner version 2.1 doesn't work with several newer Garmin handhelds. An alternative is to use a commercial product called GPSGate. You can find more information about it at `http://franson.com/gpsgate`.

Training Center

Training Center is a free log and analysis program that's compatible with Garmin fitness GPS products. It allows you to store, track, and analyze workout data.

WebUpdater

The free WebUpdater program determines which firmware version your Garmin GPS unit is running, connects to the Internet and checks if a newer version is available, and then downloads and installs the firmware for you. Simple, screen-based instructions make the process easy.

xImage

xImage is a free Windows program that saves a bitmap image of what's appearing on your Garmin GPS receiver's screen. Use a USB cable to connect your GPS receiver to a PC and then run xImage. A bitmap image is captured and saved to your hard drive. Graphics files can also be uploaded to some GPS models with xImage.

Scouting Magellan Software

Magellan (www.magellangps.com) is the second largest GPS manufacturer and offers the eXplorist and Triton lines of handheld GPS receivers.

Magellan is in the process of restructuring its line of software products. I suggest you check the company's Web site to get the most current information on available programs and maps.

Software products that you can use with Magellan mapping GPS receivers include the following.

VantagePoint

VantagePoint is Magellan's new GPS map interface program. It allows you to view compatible maps on a PC and download them to your eXplorist or Triton GPS unit. You can also manage waypoints, routes, and tracks. VantagePoint is available for download after you register for a free user account on the Magellan Web site.

Magellan discontinued much of its MapSend line of software in January 2008, including the DirectRoute street navigation program. It remains to be seen whether the company will offer updated software that works with its older Meridian, SpoTrak, and eXplorist models.

Magellan offers versions of MapSend Lite (for interfacing MapSend maps) and MapSend Manager (a waypoint and route manager) for older Magellan GPS models. Both Windows programs are free and can be downloaded from the company's Web site. Neither of these programs comes with map data.

National Geographic TOPO!

Instead of offering proprietary maps for its Triton series of handheld GPS receivers, Magellan built in compatibility with National Geographic TOPO! products. If you own a Triton, you can purchase a copy of TOPO! (1:24,000 scale U.S. maps available for each state) or Weekend Explorer 3D (regional maps based around popular recreation areas) and then select parts of maps to transfer to your GPS unit. The software displays streets and roads, but doesn't support autorouting. For more information, see www.natgeomaps.com/triton.

If you're a Magellan Triton owner, be sure to check out www.tritonforum.com for lots of information on maps and using the line of GPS devices.

Magellan map products

- **AccuTerra USA:** Downloadable regional topographic maps for the United States.

- **Topo Canada:** Downloadable Canadian regional topographic maps.

- **Topo Norway:** Downloadable topographic maps of Norway divided into regions.

- **Topo Mexico:** Topographic maps of Mexico available on a memory card.

- **StoneMaps – Amusement Parks U.S.:** Ever get lost at an amusement park? Next time download either the California or Florida collection of these maps on your GPS unit. (These same maps are available for Garmin receivers at www.stonemaps.com.)

- **Mountain Dynamics Snow Ranger:** Ski resort maps with detailed information on lifts, runs, and lodges. U.S. and European regional versions available for download. (These same maps are available for Garmin receivers at www.mountaindynamics.com.)

- **Lakes USA:** United States popular fishing spots, reservoirs, lakes, and waterways. North, South, and West regions are available either on memory cards or for download.

- **BlueNav Local Charts:** These are downloadable, international nautical charts with detailed coverage areas and enhanced marine navigation information, such as currents, tides, and port services.

- **BlueNav XL3 Charts:** Memory card nautical charts based on Navionics Gold chart data. Charts are available for North American and European regions.

If you have a Magellan GPS receiver and are interested in creating your own maps, check out the unofficial Magellan GPS tools site at www.msh-tools.com. You'll find lots of interesting programs and how-to information there.

Discovering Lowrance Software

Lowrance (www.lowrance.com) was the first manufacturer to offer a GPS receiver with uploadable maps. (For you trivia buffs, it was the GlobalMap100, which came out in April 1998.) The company continues the electronic map tradition with its line of handheld iFinder and automotive GPS receivers.

If you own a Lowrance GPS receiver or are considering purchasing one, here are the map products you can use (Lowrance also produces recreational marine electronics, and you'll definitely see a fishing and outdoors orientation with their map products):

- ✔ **MapCreate:** MapCreate is Lowrance's map program for creating GPS receiver maps of the United States (a version for Canada is also available). MapCreate USA Topo comes on a DVD with a large amount of map data, including lake depth contours, public hunting lands, coastal and inland navigation aids, and other useful features for the outdoor recreationist. Roads and streets are shown, but autorouting isn't supported. One nice feature of MapCreate is that you can select polygonal map regions to upload, which maximizes map storage space. A MapCreate screen is shown in Figure 11-2.

- ✔ **FreedomMaps:** In addition to MapCreate, Lowrance also offers the FreedomMaps line — regional maps of the United States, Europe, and Canada loaded on to memory cards. (An enhanced world basemap is also available.) The U.S. maps contain the same data found on the MapCreate product.

- ✔ **LakeMaster Pro:** High-definition, lake and pond maps for Michigan, Minnesota, and Wisconsin. These maps are sold on separate memory cards.

- ✔ **Fishing Hot Spots Pro:** Detailed maps and information on the top inland fishing lakes in the United States. Four regions are available, each sold on a preprogrammed memory card.

If you're into fishing, check out Lowrance's Enhanced Lake Maps. The free maps cover some of the most popular fishing lakes and reservoirs in the US. Just download an individual lake from the Lowrance Web site, copy it to a memory card, plug the card into a compatible GPS, and then grab some bait and your rod and go.

- ✔ **NauticPath:** Lowrance's in-house brand of offshore and coastal charts (on memory cards) for boaters. United States and international regions are available.

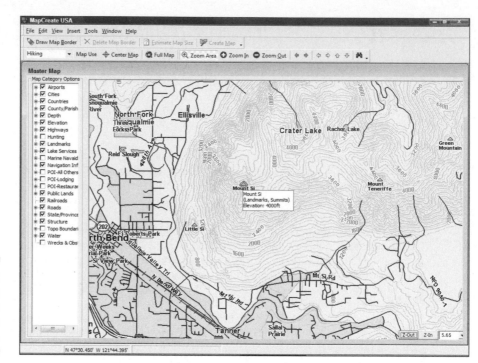

Figure 11-2:
A Lowrance
MapCreate
topographic
map.

A number of Lowrance models are also compatible with Navionics (see www.navionics.com) brand Gold and Classic memory card charts. The NauticPath charts are quite good, but the more expensive Navionics products provide even more detail.

If you own an XOG cross-over GPS unit, Lowrance has a Web site called MapSelect (www.mapselect.com) where you can download satellite and aerial images (2-meter resolution), 1:24,000 scale eTopo maps (scanned and enhanced USGS topographic maps), and 1:100,000 scale Bureau of Land Management maps. Individual maps cover approximately a 4-square mile area, the BLM maps are roughly twice the size and are priced at $5.00 per map.

In addition to maps, Lowrance also offers two free GPS utility programs.

✔ **GPXToUSR:** A utility that allows you to convert GPX format files to Lowrance's USR format. This program is especially useful for geocachers or anyone who wants to import waypoints from another manufacturer's GPS unit.

✔ **GPS Data Manager:** A reduced feature version of MapCreate designed to exchange data between a GPS receiver and a PC, including waypoints, routes, and *trails* (Lowrance's term for GPS tracks).

Navigating DeLorme Software

Unlike other GPS manufacturers that started with GPS products and then got into electronic maps, DeLorme (www.delorme.com) was providing commercial maps and gazetteers long before GPS became available to consumers. In addition to traditional maps, the company has kept up with the times, introducing the Earthmate mouse GPS unit in the mid-1990s and its first handheld GPS receiver, the PN-20, in 2007.

DeLorme has an interesting approach to its GPS maps. Instead of selling add-on map software like other manufacturers, the company ships handheld GPS units with a DVD containing its Topo USA program.

As you might expect from a company that specializes in maps, Topo USA comes with top-notch, detailed topographic vector maps covering the entire United States; including streets and roads. (Both the PC software and DeLorme's GPS receiver can autoroute.) I discuss Topo USA's PC features in Chapter 14.

Transferring data to your handheld DeLorme GPS receiver works like other GPS map programs. Select parts of the map you're interested in, save them in a receiver compatible format, and then copy the map to the receiver's memory card.

In addition to maps that come with Topo USA, the program has a feature called NetLink. NetLink allows you to purchase and download other types of maps to your PC for use with Topo USA and your DeLorme handheld receiver. Maps and their costs include

- **USGS 1:24,000 scale scanned maps** — $1 per square kilometer (DeLorme also sells a DVD with all of a single state's topo maps for $99.)

- **DOQQ (Digital Orthophoto Quarter-Quadrangle) color aerial images** — $1 per square kilometer

- **NOAA scanned nautical charts** — $10 a chart

- **SAT 10 color satellite images** — 5 cents per square kilometer (10-meter resolution)

- **USGS DOQQ black-and-white aerial images** — 10 cents a square kilometer

- **USGS Urban Areas high-resolution, color aerial images** — $2 per square kilometer

You can view sample maps and coverage areas so you know what data you can get and where.

When you click on the NetLink tab, a series of grids appears over the map you're viewing. Choose the type of map you're interested in and then click a grid for the area you want to download. A text box keeps track of the cost of the maps and gives you an idea of how long the data will take to download. (Figure 11-3 shows a sample NetLink session.)

Pay for the maps with a credit card and follow the download instructions. After you've downloaded the maps, they appear in Topo USA and you can transfer selections to your DeLorme handheld GPS receiver.

In addition to Topo USA and downloadable maps, DeLorme also offers XMap Pro. This software package was originally developed for GIS use, but the company added some features so the average user can create maps that can be loaded onto a handheld DeLorme receiver.

Although it's possible to create maps for Garmin and Magellan GPS units, this is because hackers reverse-engineered proprietary map formats. Neither Garmin nor Magellan offer software that allows an end-user to create custom GPS receiver maps.

Figure 11-3:
DeLorme
NetLink map
download-
ing feature.

To make a custom map, you have three options:

- ✔ **Use GIS data:** Because XMap is primarily a GIS tool, you can build a map with freely available, standard format GIS data, such as elevations, water features, populated places, and more.

- ✔ **Use a georeferenced map:** If you have a map with georeferenced data in GeoTIFF or MrSID format, simply import the map file and then select the area you want to transfer to your handheld. A large number of free maps and charts are available on the Internet in GeoTIFF format. I list some sources in Chapter 20.

- ✔ **Georeference your own map:** Using a graphics file, such as a scanned paper map, associate a series of points on the map with known GPS coordinates. XMap can then georeference the image, so geographic coordinates are available for the entire map. (I discuss georeferencing with OziExplorer in Chapter 16, and although the program is different from XMap, the general concepts are the same.)

XMap has the same interface as Topo USA and other DeLorme products. The process of creating and transferring maps is straightforward.

To find out more about DeLorme's handheld GPS units and map products, visit the company's forums at http://forum.delorme.com.

Chapter 12

Finding Places and Coordinates

• •

In This Chapter

▶ Using online gazetteers

▶ Converting coordinates

• •

*W*hen you start using a handheld GPS receiver, you soon discover that it's pretty numbers-oriented. There's time, speed, distance, altitude, and (of course) the location coordinates. But quite often, you want names to go with those numbers. Or you might need to convert those coordinate numbers into another format. That's where this chapter comes in. Read on to discover how to locate places by their names and get their coordinates (and other information) and how to easily convert coordinates from one coordinate system to another.

Finding Your Way with Online Gazetteers

Sometimes you need a little bit more information about a location:

- ✔ You know a place name, but you don't know exactly where the place is located.

- ✔ You've heard about a place but don't know whether it's a mountain peak, a river, or a town.

- ✔ You generally know where a place is, but you need the exact latitude and longitude or Universal Transverse Mercator (UTM) coordinates.

In these cases, you can turn to a *gazetteer,* which is a collection of place names with such useful data as geographic coordinates, elevation, and the type of place it is. Gazetteers are usually published as books, but digital versions are also available. The U.S. government has two free online gazetteer services:

- ✔ **GNIS** provides information about places in the United States.

- ✔ **GNS** has information about locations all over the world.

Using the Geographic Names Information System (GNIS)

The Geographic Names Information System (GNIS) is the federal repository of geographic name information. The database contains information on nearly 2 million physical and cultural geographic features in the United States and its territories, including cities, dams, islands, schools, and other designated feature types. You can search for feature information at the GNIS Web site: `http://geonames.usgs.gov/pls/gnispublic`.

The GNIS search page has a number of different data fields (as shown in Figure 12-1) that you can use to narrow down your search, including

- ✓ **Feature Name:** The name of the feature you're looking for. This can be either the whole name or a part of the name.

  Feature name searches aren't case sensitive.

- ✓ **Exact match:** Check this box to search for a feature with the exact name you enter.

- ✓ **Exclude variants:** Some features have other names in addition to their primary name. If you check this box, records with matched variant names are excluded.

- ✓ **State:** From this drop-down list, select the state or territory where the feature is located.

- ✓ **County Name:** If you click the County Name button, a drop-down list box shows all the counties in the selected state.

  If you know the county where the feature is located, enter it to speed up your search.

- ✓ **Feature Class:** The Feature Class drop-down list contains all the feature types, such as bridges, canals, lakes, and populated places. If you know what the feature is, select its type.

- ✓ **Elevation:** The Elevation drop-down list lets you select an operator (between, equals, higher than, or less than). Enter the elevation to the right of the drop-down list and select the feet or meters radio button below it.

Follow these steps to perform a basic search for a feature:

1. **Go to the GNIS Web site at** `http://geonames.usgs.gov/pls/gnispublic`.

2. **Enter the feature name that you want to search for in the Feature Name text box.**

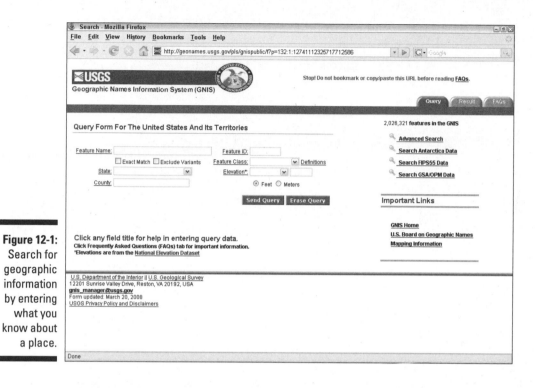

Figure 12-1:
Search for geographic information by entering what you know about a place.

3. **From the State drop-down list, select the state where the feature is located.**

4. **Click the Send Query button.**

The more you narrow a search, the faster it is. For example, if you know the county where a feature is located, select it. If you don't know much information about the feature, be patient. The GNIS server can be pretty slow.

If GNIS finds any records that match your search criteria, it lists all the matching features. Figure 12-2 shows that a search for *Horse Butte* found three matches. Information about the features includes

- ✔ Feature name
- ✔ State
- ✔ ID (FIPS 55 code for U.S. locations)
- ✔ Class
- ✔ County
- ✔ Latitude

> ✔ Longitude
>
> ✔ Map (USGS 7.5 minute map that the feature appears on)
>
> ✔ Elevation
>
> ✔ Date entered

If you aren't sure which result you need, use this information to narrow your search.

After you pick a feature, click its Feature Name link to display more information. A new page (as shown in Figure 12-3) displays additional information. You already saw most of this data in the previous screen, but on the right side of the window is a list of Mapping Services for displaying the feature on a number of different online map sites. Simply click one of the services to open a new window with the map of your choice.

You need an Internet connection to access the GNIS database. In addition to the online search capabilities at the GNIS Web site, you can download text files of all the features and associated information for each state. The files are quote- and comma-delimited and can be opened with your database or spreadsheet. The files come in compressed and uncompressed formats; if you have a slow Internet connection, download the Zip files.

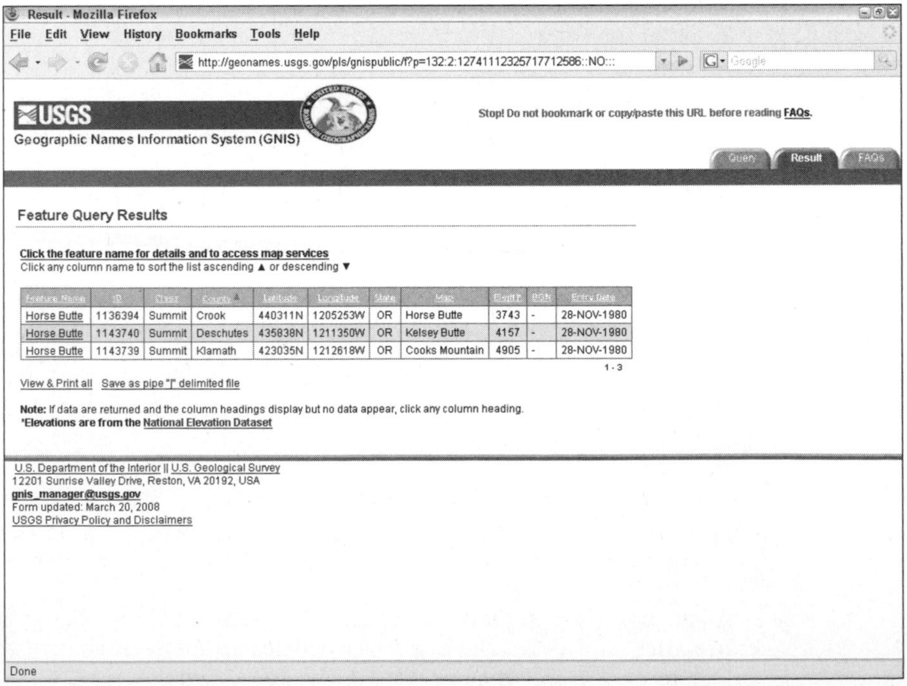

Result - Mozilla Firefox

File   Edit   View   History   Bookmarks   Tools   Help

http://geonames.usgs.gov/pls/gnispublic/f?p=132:2:12741112325717712586::NO:::

≊USGS
Geographic Names Information System (GNIS)

Stop! Do not bookmark or copy/paste this URL before reading **FAQs**.

Query   Result   FAQs

Feature Query Results

Click the feature name for details and to access map services
Click any column name to sort the list ascending ▲ or descending ▼

Feature Name	ID	Class	County	Latitude	Longitude	State	Map	Elev Ft	BGN	Entry Date
Horse Butte	1136394	Summit	Crook	440311N	1205253W	OR	Horse Butte	3743	-	28-NOV-1980
Horse Butte	1143740	Summit	Deschutes	435838N	1211350W	OR	Kelsey Butte	4157	-	28-NOV-1980
Horse Butte	1143739	Summit	Klamath	423035N	1212618W	OR	Cooks Mountain	4905	-	28-NOV-1980

1 - 3

View & Print all   Save as pipe "|" delimited file

Note: If data are returned and the column headings display but no data appear, click any column heading.
*Elevations are from the National Elevation Dataset

U.S. Department of the Interior || U.S. Geological Survey
12201 Sunrise Valley Drive, Reston, VA 20192, USA
gnis_manager@usgs.gov
Form updated: March 20, 2008
USGS Privacy Policy and Disclaimers

Done

Figure 12-2:
GNIS
search
display
results.

![Detail - Mozilla Firefox browser window showing the USGS Geographic Names Information System (GNIS) Feature Detail Report]

Feature Detail Report

Feature ID: **1143740**
Name: **Horse Butte**
Class: **Summit**
Citation: **Collected during Phase I data compilation (1976-1981), primarily from U.S. Geological Survey 1:24,000-scale topographic maps (or 1:25K, Puerto Rico 1:20K), various edition dates, and from U.S. Board on Geographic Names files.**
Entry Date: **28-Nov-1980**
Elevation(ft/m)*: **4157/1267**

Counties

Sequence	County	Code	State	Code	Country
1	Deschutes	017	Oregon	41	US

Coordinates (One point per USGS topographic map containing the feature, NAD83)

Sequence	Latitude(DEC)	Longitude(DEC)	Latitude(DMS)	Longitude(DMS)	Map Name
1	43.9773402	-121.2305829	435838N	1211350W	Kelsey Butte

Mapping Services
Click the link to display the feature in U.S. mapping services.

USGS The National Map
HomeTownLocator
TopoZone.com
GNIS in Google Map
Microsoft Virtual Earth
TerraFly.com
TerraServer DOQ
TerraServer DRG
Find the Watershed
MapQuest
Yahoo! Local Maps
Expedia

Figure 12-3: Detailed information from the GNIS database.

Ever wanted to name a mountain or another land feature after yourself or a loved one? The U.S. Board on Geographic Names is responsible for naming and renaming features that appear on USGS maps. If there's an unnamed geographic feature, you can propose a new name for it (or suggest a name change for an existing feature). The Board even has an online form that you can fill out. Submitting the form doesn't guarantee you'll automatically get some peak named after Uncle Harry; this is a rather big deal. For instructions on filling out the online form, see http://geonames.usgs.gov/dgnp/dgnp.html.

Using the GEOnet Name Server (GNS)

The GEONet Name Server (GNS) searches for features around the world. GNS is located at http://earth-info.nima.mil/gns/html/index.html.

The database contains over 3.5 million features and over 5 million place names for locations outside of the United States. The military relies on GNS for its operations, so the database is updated every other week.

GNS is primarily designed for military use. Some of the search criteria and information in the database isn't very useful to the average civilian. At best, expect to find these types of information for a given feature:

> ✔ Country
>
> ✔ Type
>
> ✔ Latitude and longitude

To perform a basic GNS search for a feature, follow these steps:

1. **Go to the GNS Web site at `http://earth-info.nga.mil/gns/html/index.html`.**

2. **Click the GNS Search link in the GNS Main Menu at the left of the page.**

   The GNS search page is shown in Figure 12-4.

   You can enter many different types of search criteria, but for now, only do a name search.

3. **In the Name text box, enter the name of the feature you're looking for.**

   A drop-down list to the right of the text box lets you narrow your search with these options:

   • *Starts With*

   • *Is an Exact Match*

   • *Contains*

   • *Ends With*

   • *Fuzzy Search*

4. **If you know the country the feature is located in, select the country from the list.**

5. **Click the Search Database button.**

Finding street address coordinates

Geocoding is the process of determining latitude and longitude from a street address. There are a number of Web-based geocoding services, and here's an example of how to use one of them.

1. **Go to `http://geocoder.us` (for the United States) or `https://geocoder.ca` (for Canada) and enter a street address.**

2. **Click the Search button (or GeoCode it on the Canadian version).**

Simple as that. A new window opens with the latitude and longitude, and the street address plotted on a map. The accuracy of this information varies, but usually it puts you in the general vicinity of the address you want. If you really want precise information, visit the address with a GPS receiver and record the coordinates.

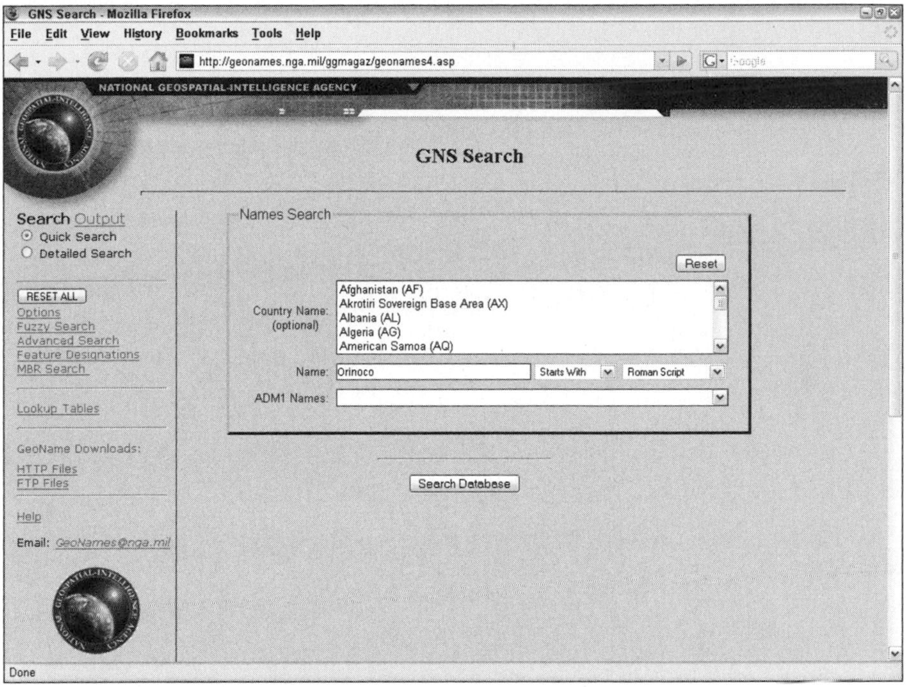

Figure 12-4:
The GNS
search
page.

For GNS advanced searches, you can enter an extensive list of feature types as part of your search criteria. These include oil pipelines, refugee camps, and vegetation types. Other options limit searches by the latitude and longitude boundaries of a rectangle, use special character sets from foreign languages, and use government and military codes as part of the search. If you want to run these advanced searches, the GNS Web site has links with explanations.

If GNS finds records that match your search criteria, the features appear on a new page. A number of pieces of information are displayed (as shown in Figure 12-5). For the average civilian user, the most useful data includes

- ✔ **Name** of the feature.

- ✔ **Region** of the world where the feature is located. Click the link to get an explanation of the abbreviation.

- ✔ **Designation:** The feature type (such as populated locality, farm, or reservoir). Click the link to get the definition of the abbreviation.

- ✔ **Latitude and longitude** coordinates.

- ✔ **Area:** Country and state or province information for the feature. Click the link for the meaning of the code.

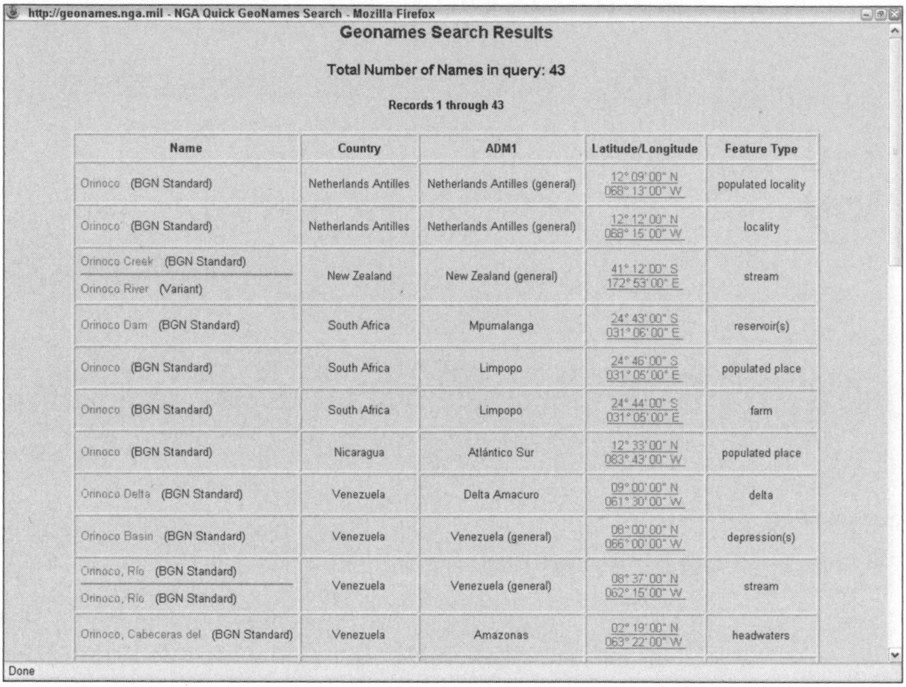

Name	Country	ADM1	Latitude/Longitude	Feature Type
Orinoco (BGN Standard)	Netherlands Antilles	Netherlands Antilles (general)	12° 09' 00" N 068° 13' 00" W	populated locality
Orinoco (BGN Standard)	Netherlands Antilles	Netherlands Antilles (general)	12° 12' 00" N 068° 15' 00" W	locality
Orinoco Creek (BGN Standard) Orinoco River (Variant)	New Zealand	New Zealand (general)	41° 12' 00" S 172° 53' 00" E	stream
Orinoco Dam (BGN Standard)	South Africa	Mpumalanga	24° 43' 00" S 031° 06' 00" E	reservoir(s)
Orinoco (BGN Standard)	South Africa	Limpopo	24° 46' 00" S 031° 05' 00" E	populated place
Orinoco (BGN Standard)	South Africa	Limpopo	24° 44' 00" S 031° 05' 00" E	farm
Orinoco (BGN Standard)	Nicaragua	Atlántico Sur	12° 33' 00" N 083° 43' 00" W	populated place
Orinoco Delta (BGN Standard)	Venezuela	Delta Amacuro	09° 00' 00" N 061° 30' 00" W	delta
Orinoco Basin (BGN Standard)	Venezuela	Venezuela (general)	08° 00' 00" N 066° 00' 00" W	depression(s)
Orinoco, Río (BGN Standard) Orinoco, Río (BGN Standard)	Venezuela	Venezuela (general)	08° 37' 00" N 062° 15' 00" W	stream
Orinoco, Cabeceras del (BGN Standard)	Venezuela	Amazonas	02° 19' 00" N 063° 22' 00" W	headwaters

Figure 12-5: The results of a GNS search for *Orinoco.*

Waypoint repositories

A number of Web sites provide waypoints that you can upload to your GPS receiver. (A *waypoint* is a set of coordinates for a location.) With these waypoint repositories, you can search a region or for a feature name for its waypoints in a database. If a waypoint has been logged in the site database, you can download the waypoint and then upload the coordinates to your GPS receiver. Some of the waypoint sites include

- `http://wayhoo.com`: This site converts GNIS feature information into waypoints. There are also coordinates for airports and a database where users can upload waypoints.
- `www.travelbygps.com`: This site holds a collection of waypoints for interesting places all over the world, including photos and descriptions. The site has an extensive collection of links to Web sites with special interest waypoints.
- `www.trailguru.com`: A collection of user-submitted waypoints, tracks and routes for off-road adventures. The site features a Google Maps interface that allows you to zoom in on a location and check if any nearby GPS data has been posted.
- `http://www.swopnet.com/waypoints/`: A large collection of links to other waypoint sites, organized by location.

You can also download tab-delimited text files from the GNS Web site for any country that contain features and information. This data can easily be imported into spreadsheets and databases.

Although GNS is pretty cool (and authorative), its user interface can be a bit daunting; at times the server is downright pokey. If I need a quick set of coordinates for a location outside the U.S., I often turn to either the World Gazetteer (www.world-gazetteer.com) or the Global Gazetteer (www.fallingrain.com/world/index.html). Both of these sites let you find place information quickly and easily.

Converting Coordinates

Sometimes you might need to convert from one coordinate system to another. For example, you might have the coordinates of a location in latitude and longitude and a map that has only a UTM grid. Although there are algorithms that you can use to convert data with pencil and paper or a calculator, let conversion utilities do the hard work for you.

The Graphical Locator

The Environmental Statistics Group at Montana State University hosts a very powerful online tool called the Graphical Locator. It's a cross between a gazetteer and a coordinate utility converter. Some of its features are

- A map of the United States that shows coordinate information when you click a location. You can zoom in on regional and state maps. (The maps only show geographic features, not feature names.)

- Coordinate conversion utilities for latitude and longitude, UTM, and township, range, and section.

- Extensive information on selected locations, including latitude and longitude; township, range, and section; UTM; elevation; state

and country; nearest named features and distances; and USGS 7.5 minute topographic map name.

The Graphical Locater is designed to work with locations within the United States. It's fairly easy to use; its author, D.L. Gustafson, has extensive online documentation on the utility.

I use the Graphical Locator for quickly getting rough latitude and longitude coordinates of a location. If there are no place names on an online map, I consult a paper map with place names to zero in where I want to get location information. This works well if you're unfamiliar with the terrain of an area.

To check out the Graphical Locator, visit www.esg.montana.edu/gl.

Using GeoTrans

GeoTrans is a popular, free Windows program developed by the Department of Defense (DoD). (GeoTrans is shown in Figure 12-6.) You can convert coordinates from many coordinate systems and datums. GeoTrans is available for download at `http://earth-info.nga.mil/GandG/geotrans`.

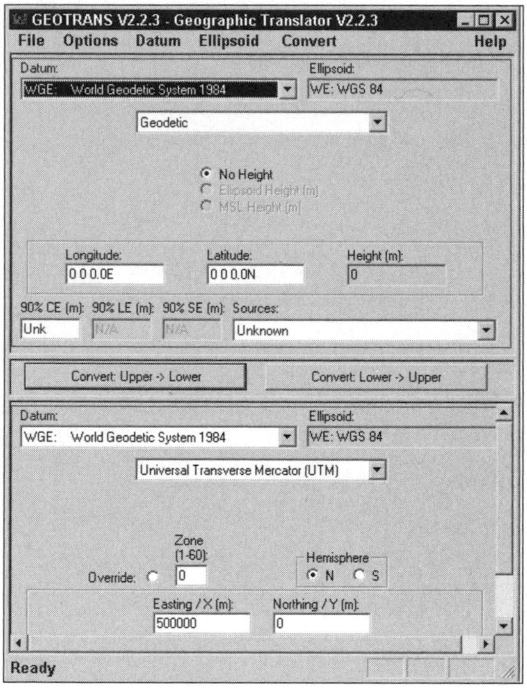

Figure 12-6:
GeoTrans
converts
coordinates
to another
format.

Follow these steps to convert between coordinate systems with GeoTrans:

1. **Select the map datum used with the coordinates from the drop-down Datum list.**

2. **Select the coordinate system from the drop-down list below the datum.**

   Use Geodetic if you're converting from latitude and longitude.

3. **Enter the coordinates in the appropriate text boxes.**

   • If you've converting from latitude and longitude, enter the coordinates.

   • If you're converting from UTM, enter the Zone, Hemisphere, Northing, and Easting.

4. **In the lower part of the window, select the datum and the coordinate system that you want to convert to.**

5. **Click the Convert Upper⇨Lower button.**

   The converted coordinates appear in the lower part of the window.

GeoTrans' Help file covers these advanced features if you need them:

✔ Converting datums used in foreign countries

✔ Determining distance errors when converting between maps and scales

Using online conversion utilities

If you don't need to convert coordinates on a regular basis, you can save some time and hard drive space by using a Web-based coordinate conversion utility instead of installing GeoTrans. Most conversion sites are pretty straightforward to use; just enter the coordinate values that you want to convert and click a button. These sites are a few of the most popular:

✔ **For simple latitude and longitude conversions,** see http://myweb.polyu. edu.hk/~04329143d/Location.htm.

✔ **For latitude and longitude, UTM, and Township and Range conversions,** go to www.esg.montana.edu/gl.

✔ **For advanced online and standalone conversion tools,** visit the U.S. National Geodetic Survey at www.ngs.noaa.gov/TOOLS.

Chapter 13

On the Road with DeLorme Street Atlas USA

*I*n the pre-PC days, taking a trip across town, a state, or the country to visit some place you'd never been before often involved planning worthy of a major expedition. You'd have to carefully check maps, trying to figure out the shortest and fastest routes, guessing when and where you'd need to stop for gas, scribbling down notes, and highlighting roads on paper maps.

That's all changed with free street map Web sites (which I discuss in Chapter 18) and inexpensive, easy-to-use street navigation software.

The Web sites are handy, but if you don't have an Internet connection or are looking for more mapping features, you should investigate street navigation software. Just run a program on your PC and enter the address of your starting point and the final destination. Then, a few mouse clicks later, you have both a map and exact turn-by-turn directions for how to get from Point A to Point B.

And as an added bonus, if you have a laptop and GPS receiver, you can take this software on the road with you, track your location in real time, and get helpful hints in reaching your destination. (Most street navigation programs also have versions that run on PDAs for ultimate portability.) Several street navigation software packages are on the market that can keep you from getting lost. They all generally work the same, with the primary differences in the user interface and support of advanced features.

If you've never used a street navigation program before, this chapter gets you moving in the right direction. I focus on DeLorme's Street Atlas USA, showing you its basic features and how to use them. I then list other available street navigation programs.

Discovering Street Atlas USA Features

Street Atlas USA displays road maps of the United States, Canada, and Mexico, finds addresses, and creates routes between two or more points. Check out a few other program features that are important to know about.

- **POIs:** All street navigation programs contain extensive databases of POIs. POIs refer to *Points of Interest*, not to Hawaiian side dishes made from taro root. POIs include restaurants (some of which might serve poi), hotels, parks, gas stations, and other locations you might be interested in while traveling. Street Atlas USA has a POI database that contains over four million businesses, services, and organizations.

- **Voice support:** If you're using a laptop and GPS receiver as part of a car navigation system, Street Atlas USA can give you voice instructions when you need to turn to reach your destination. You can also use a voice recognition feature to give Street Atlas USA commands instead of using a keyboard or mouse.

- **Routable roads:** A big issue that all map companies face is ensuring that their road data is accurate, which can be very challenging considering the number of new roads that are built every year. Street Atlas USA has a feature that allows you to draw in roads that are missing on a map. After you create a road, Street Atlas USA can use it when calculating routes.

- **Mini-laptop interface:** If you're using a Windows mini-laptop (say, an Asus Eee PC), Street Atlas USA offers a UMPC (Ultra Mobile PC) mode during installation that configures its interface for the smaller screen size.

- **Aerial photos:** You can purchase and download aerial photos and then overlay streets and POI information on the birds-eye view.

- **Customizable maps:** Street Atlas USA has an extensive collection of drawing tools for customizing maps with symbols, shapes, and text annotations.

Street Atlas USA has many more features than I can cover in the space of this chapter (such as support for Palm and Pocket PC devices, downloadable maps, and measuring distances and trip planning that takes fuel consumption as well as the number of hours spent driving into consideration). To find out more about all the program's features, visit www.delorme.com.

Street Atlas USA is also available in a Plus edition that comes with 150 million U.S. and Canadian searchable business phone listings, extensive data import capabilities, and higher-end print and draw tools.

Navigating Street Atlas USA

The first thing that you notice about Street Atlas USA is that it doesn't use a familiar Windows, menu-based user interface. DeLorme uses a unique user interface with its mapping programs; after you get the hang of it, it's pretty easy to use. I walk you through the user interface and then show you how to move around inside a map.

Be sure to quickly browse through the PDF user guide and help file that comes with Street Atlas USA, which you can access by clicking the question mark button at the top of the window. Street Atlas USA has a rich set of commands, often offering you several different ways to perform a single task or operation.

Exploring the Street Atlas USA interface

The Street Atlas USA user interface is made of four different parts (as shown in Figure 13-1). They include the

- ✔ **Map:** The main map takes up most of the screen and is where all the action takes place. You'll find roads, bodies of water, parks, businesses and services, and other features displayed.

- ✔ **Control Panel:** The Control Panel, located to the right of the map, contains commands for moving around in the map and zooming in and out.

- ✔ **Tab functions and options:** Primary mapping commands and options are underneath the map in a series of tab items. For example, click the Find tab for searching commands and options.

You can shrink the size of the Tab area to show more map area by clicking the down-arrow icon in the right corner of the Tab area. If the Tab area is minimized, click any tab to expand it.

- ✔ **Overview Map:** The Overview Map appears to the right of the Tab area and contains a small map with a larger overview of the main map that you're viewing.

You can customize the Tab area by clicking the Options button in the window title and choosing the Tab Manager menu item. Use the Tab Manager to show, hide, or reorder the tabs.

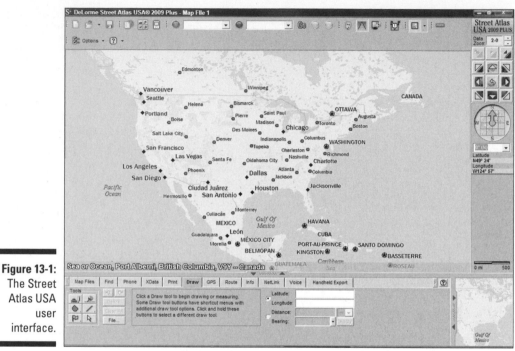

Figure 13-1:
The Street
Atlas USA
user
interface.

Zooming in and out

With Street Atlas USA, you can view the entire United States and then zoom in for street-level detail. As you zoom in, the Data Zoom level, which displays at the top of the Control Panel, increases. Data Zoom level 2–0 shows the entire United States, and Data Zoom level 16–0 shows the maximum amount of detail for a location.

Beneath the Data Zoom are three buttons that control zooming. These buttons, each with red arrows and pictures of the Earth, are from left to right

- ✔ **Zoom out three levels:** Click the button with three arrows pointing away from Earth.

- ✔ **Zoom out:** Click the button with the red arrow pointing away from Earth.

- ✔ **Zoom in:** Click the button with the red arrow pointing toward Earth.

In addition to the zoom buttons, DeLorme also uses *Octave* controls, which are up and down arrows next to the Zoom Data level value, allowing you to have finer control over zooming in and out. Click the up arrow to zoom out; click the down arrow to zoom in.

When you click an Octave control, notice that the Zoom Data level number changes. The number to the right of the dash next to the Zoom Data level is the octave value. For example, if the current Zoom Data level were 12-3, clicking the Octave down arrow would zoom in and change the value to 12-4. (*Octaves* range between 0 and 7, just like a diatonic music scale.)

You can also zoom in on a specific area by holding down the left mouse button and dragging down and to the right. This draws a rectangle and zooms in to that area when you release the mouse button. You can zoom out by holding down the left mouse button and dragging up and to the left.

Moving around in Street Atlas USA

Mouse around a little inside a Street Atlas USA map. Notice that while you move the cursor around, information appears on the lower edge of the map. Anytime you move the cursor over a map feature, whether it's a road, river, or even some open space, a line of text appears at the bottom of the map with a brief description of the feature.

The numbers in parentheses that appear before a street name description show the range of street addresses in the general vicinity of the cursor. This is handy for getting a quick idea of addresses on a particular street.

You're probably going to want to see more map than what appears on the screen, and Street Atlas USA has several ways to move the map, including

- ✔ **Centering:** Click a location to center the map over the cursor.
- ✔ **Dragging:** Whenever you move the cursor to the edge of the map, it turns into a hand icon. Hold the left mouse button down and drag the map to scroll.
- ✔ **Arrow keys:** You can use the keyboard arrow keys to move the map in the direction of the arrow key you pressed.
- ✔ **Compass Rose:** In the Control Panel, beneath the zoom tools, is the Compass Rose. This is a series of nine buttons with yellow arrows overlaid on a picture of a globe. Click a button to scroll the map in the direction of the arrow.

If you click the middle button in the Compass Rose, the previously viewed map displays. You can view the last 256 previously displayed maps by clicking this button.

Getting POI information

Getting Point of Interest information from Street Atlas USA is a snap. Here's how.

When you zoom in to level 15, you start to see POIs on the map, such as restaurants, gas stations, theaters, hotels, and other businesses and services. Symbols appear that show you what the POI is. For example, a plate with a fork and knife means a restaurant.

If you don't know what a symbol means, click the question mark button in the window title and choose Map Legend from the menu to display a list of all the map symbols and their meanings.

When you move the cursor over a POI, the business or service name, its phone number (if available), and the type of POI are shown at the bottom of the map. You can also get more information about a POI (or any map feature) by moving the cursor over the POI, right-clicking, and choosing Info from the pop-up menu. Information about the POI appears in the Tab area, which you can view by clicking the Info tab, as shown in Figure 13-2.

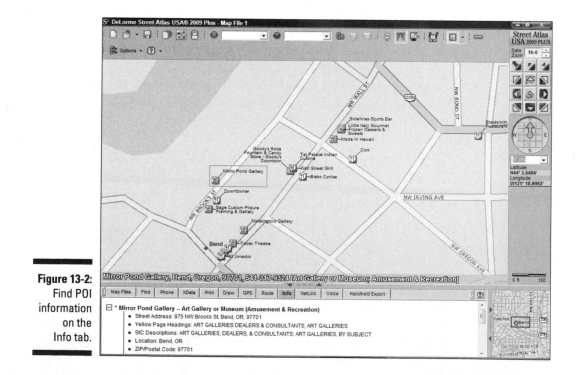

Figure 13-2:
Find POI
information
on the
Info tab.

POI databases aren't perfect. If you look up restaurants, gas stations, and other businesses in your city, you'll probably find a few listed that are out of business and others that are missing. DeLorme and other map companies try to keep POI data as current as possible, but because businesses come and go so often, it's difficult to keep up with all the changes.

Finding an Address with Street Atlas USA

Whenever I'm in San Francisco, I always try to visit Sam Wo's, my favorite authentic, Chinese noodle joint, located smack-dab in the middle of Chinatown. Anytime I hear about friends or acquaintances heading to San Francisco, I always tell them to head downtown and check out this famous Chinese fixture. It's been around for 100 years, Kerouac and Ginsberg hung out there, and it was home to arguably the rudest waiter in the world.

If you don't know San Francisco, finding this restaurant can be challenging. And even if you can get around the City by the Bay fairly well, giving precise directions to visitors can be a little demanding. This is where street navigation programs like Street Atlas USA really shine. As an example of finding a location and creating a map to get to it, here's how to find Sam Wo's restaurant:

1. **Click the Find tab and then click the QuickSearch button.**

   You can perform several different types of searches, but this is a simple QuickSearch. (I talk about Advanced searches and GPS Radar in a minute.) The Find tab options and commands are shown in Figure 13-3.

Complete campgrounds

Street Atlas USA is popular with RV owners who travel across the United States. With a laptop and a GPS receiver, you have a personal navigation system for vacations and touring. And if you're a male (cough), you significantly reduce the chances of getting lost and being forced to ask for directions.

A POI database is a friendly traveling companion, but if you're an RV-er, you can supplement it with even more useful information.

The Discovery Motorcoach Owners Association Web site has an extensive list of U.S. and Canada campgrounds, including Wal-Mart and Costco locations, which you can import into Street Atlas USA and other mapping programs.

To download the free campground database and get complete instructions on how to load it into Street Atlas USA, go to www.discovery owners.com/cginfo.htm.

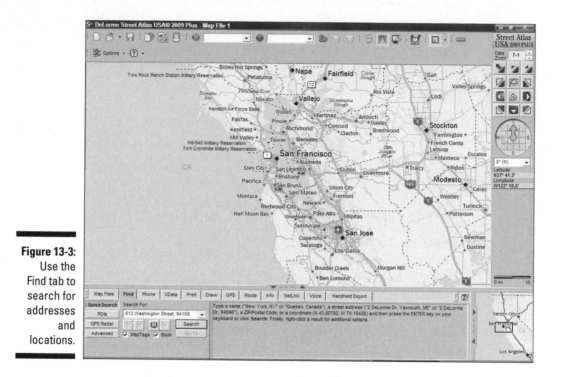

Figure 13-3:
Use the
Find tab to
search for
addresses
and
locations.

2. **In the Search For text box, enter the address you want to find.**

   Sam Wo's restaurant is located at 813 Washington Street, 94108. You can enter the city and state along with the address, but if you know the ZIP code, it's quicker and easier to use it instead.

3. **Click the Search button.**

   Street Atlas USA searches through map and street data; if the address is valid, a map of the location displays. In this case, Sam Wo's is exactly where I remember it to be. See the resulting map in Figure 13-4.

The restaurant is in the right place, but the name isn't correct. (It's Wo, not Who.) What to do? If you're connected to the Internet, Street Atlas USA has a way to tell DeLorme about map and POI errors. In the NetLink tab, click the Corrections button and let the guys at DeLorme know about an error. I did, and in future versions, I hope Sam's name is spelled correctly.

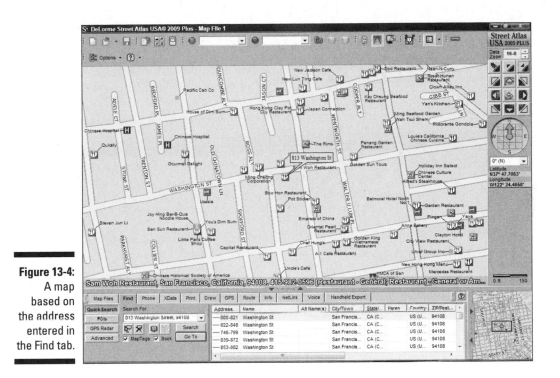

In addition to a QuickSearch, which is great for basic searches involving places, addresses, ZIP codes, and coordinates, you'll find two other options in the Find tab:

- **Advanced searches:** As the name suggests, this option performs more detailed, advanced searches. You can search by categories (such as rest areas, schools, and parks) as well as specify searches take place within a certain area (such as a ZIP code, county, or the current displayed map).

- **Radar searches:** This is cool feature that lets you search for travel-related POIs within a certain distance of the current center of the map — or if you're using a GPS receiver with Street Atlas USA, from your current location. For example, you could request, "Find me all the restaurants within one mile," and a list of nearby restaurants would be displayed. When you click a restaurant in the list, Street Atlas USA draws a route on the map showing you how to get to the restaurant and tells you the address, how far away it is, and how long it will take to arrive.

Getting from Here to There with Street Atlas USA

Knowing the address of some place and where it's located is a start, but getting there in a timely fashion without getting lost or frustrated is the true test of a driver (or his or her navigator).

One of the benefits of street navigation software is that it can automatically generate a route between two or more points. The software examines the roads between your starting point and destination, measuring distance and factoring in speed limits to select either the shortest or fastest route.

Street navigation software can give you only its best guess when it comes to a route. A program can't account for new roads that were built after the map data was compiled, road condition, bad neighborhoods, or local traffic patterns. You'll probably be able to find faster and more direct routes in cities based on your own local knowledge and experience. Although routes from software might not always be perfect, they're much more accurate than guessing or driving around aimlessly.

Street Atlas USA has a number of powerful options for creating routes. To give you a better idea of how route finding works, start simple.

Creating a route

If you've followed the chapter to this point, you can use the example map created in the earlier section, "Finding an Address with Street Atlas USA." Suppose I want to give my friends this map of Chinatown, showing where Sam Wo's restaurant is located. Great, but that's not going to be much help if they've just arrived at the airport and are driving into the city. For their needs, I use Street Atlas USA to create a route for them to follow to noodle nirvana.

1. **Click the Route tab.**

2. **In the Start text box, enter your starting address.**

   I use a cool Street Atlas USA feature that lets you use the three-letter code of an airport. I enter **SFO**, which is the code for San Francisco International Airport.

3. **For the destination, enter the address in the Finish text box.**

   Here, I enter Sam Wo's address: **813 Washington Street, 94108.**

When you enter a ZIP code as part of the search, Street Atlas USA removes the ZIP code from the Finish text box when the calculated route displays. The ZIP is required for the search but isn't shown with the route.

4. **Click the Calculate button.**

Presto! A map appears with an outlined route from the airport to Sam Wo's, as shown in Figure 13-5. The route distance is shown as well as an estimate of how long it will take to get there.

Street Atlas USA calculates routes for not only motorized vehicles, but also provides the best-suggested route for bicyclists or walkers.

Remember that your Calculate result is based on driving the speed limit with a normal amount of traffic. You can zoom in on the map to get more detail.

On all street navigation software, you'll find that the travel times for routes tend to be conservative estimates. Nine times out of ten, you end up taking less time to reach your destination.

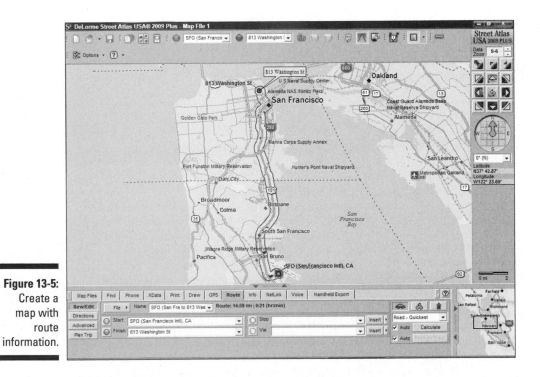

Figure 13-5:
Create a
map with
route
information.

Getting directions

Maps with highlighted routes are pretty cool, but sometimes it's nice to have a set of plain old *turn right on Main Street, then take a left at the light* directions. After you create a route (see the preceding section), Street Atlas USA can provide you with turn-by-turn directions to get to your destination.

On the Route tab, click the Directions button. You'll get a list of turn-by-turn directions with street names, distances, and total times for each part of the route. The directions information is shown in Figure 13-6.

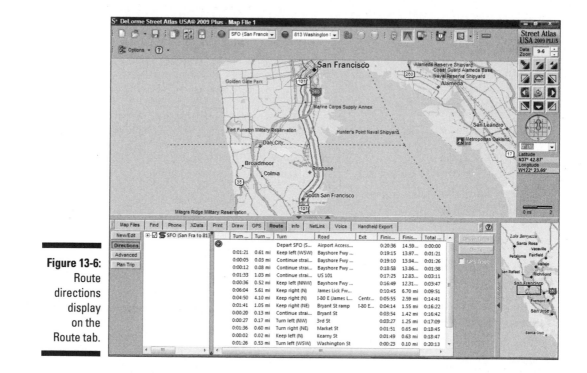

Figure 13-6:
Route
directions
display
on the
Route tab.

Printing and saving directions

After you have a map and driving directions (read earlier sections to follow me to Sam Wo's), you may need to make copies. Street Atlas USA has a number of options for getting the information off your computer screen and printing it to paper or saving it to a file.

To print or save maps and directions, click the Print tab. You see two buttons:

- ✔ Click the *Map* button for options to print or save the current map.
- ✔ Click the *Route* button for print and save options of the current route.

Saving and printing maps

All the Print and Save command buttons are next to the Overview Map on the far right of the Print tab area. You have buttons for

- ✔ Printing the current map
- ✔ Saving the print area to disk as a graphics file
- ✔ Copying the print area to the Clipboard

If you select the Print Preview check box, the map is reduced in scale, and a rectangle displays around the print area.

Printing routes

You can also print route information in a number of different formats, including an overview map with directions, turn-by-turn details, and as *s trip maps* (detailed maps that follow a route with directions in the margins). Check the type of directions you'd like to create, using the Save and Print command buttons at the far end of the Tab area to output them. Figure 13-7 shows a preview of a *Travel Package,* which is a format that prints the route map with directions at the bottom of the page.

If you want only a text version of the route directions, click the Route button in the Print tab and then mark the Directions check box. You can then either save the directions as a text file or copy them to the Clipboard and paste them into an e-mail or word processor document.

If you end up in Chinatown, the entrance to Sam Wo's restaurant is through the street-level kitchen, so don't think you're in the wrong place. Head up the narrow stairs to the second floor and don't disturb the cooks on your way up. *Bon appétit!*

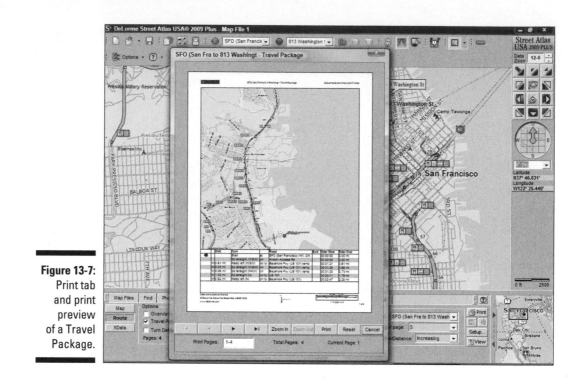

Figure 13-7:
Print tab
and print
preview
of a Travel
Package.

Moving Maps with Earthmate

In addition to its mapping software, DeLorme also offers the small Earthmate GPS receiver that's designed to work with Street Atlas USA and other map programs that use GPS data. (You can use other types of GPS receivers with Street Atlas USA, too, but the Earthmate is designed to work with the program with a minimal amount of setup.)

In 2008, DeLorme introduced the LT-40 version of the Earthmate GPS receiver. This model features a high-sensitivity GPS chip and performs better than older models.

You can connect the Earthmate to a laptop via a Universal Serial Bus (USB) cable. (DeLorme also sells a GPS receiver that has both USB and Bluetooth interfaces.) After you have the GPS receiver hooked up and Street Atlas is running, the two general modes of operation are

✔ **General navigation:** Street Atlas USA processes the received GPS data and displays your current position on the map with a series of dots that show where you've traveled. While you move, the map automatically moves to show your position. Additionally, your speed, direction of travel, and GPS satellite information is shown in the program's GPS tab.

✔ **Route:** After you create a route (see the earlier section "Creating a route") in addition to the general navigation features, Street Atlas USA displays turn-by-turn directions onscreen, informing you how far your destination is and the travel time to your destination. If you have voice enabled, Street Atlas USA announces the directions; you can choose from several different types of voices.

If you're driving by yourself, be careful when using the GPS features of Street Atlas USA with your laptop sitting next to you on the passenger seat. Fight the tendency to get distracted from your driving while you look at the screen and use the mouse and keyboard to enter commands. I recommend someone riding shotgun — that is, a navigator who's in charge of running Street Atlas USA. If you drive solo a lot, get a laptop-mounting bracket that places your computer in a more visible and easy-to-use location.

Visit the DeLorme forums at `http://forum.delorme.com` for more information about Street Atlas USA, including lots of tips, tricks, and advice from a friendly user community.

Other Street Navigation Software

In addition to Street Atlas USA, here are two other popular street navigation programs available for navigating the roads of America. I don't have enough space to describe them fully, but I offer some general information in case you're shopping for software.

New versions of street navigation software are usually released annually with new street and POI data as well as new features. If you travel in an area that doesn't experience much growth or change, you probably don't need to upgrade every year. On the other hand, because all the street navigation software packages retail between $50 and $150, it's not that expensive to stay current. (On all products, watch for frequent rebates, upgrade deals, and other discounts to get the cost even lower.)

Microsoft Streets & Trips

Microsoft's Streets & Trips is a popular alternative to DeLorme's software. The program has all the basic street navigation features, including some advanced features such as saving a map as a Web page, downloading current road construction information from the Internet, and creating drive time zones (such as *show me all the places I can drive to from a certain location in under an hour*). The software comes with an enhanced GPS receiver that provides real-time information about traffic conditions and gas prices, using Microsoft's wireless MSN Direct service (a one-year subscription is also included). If you already have a GPS and MSN Direct receiver, you can just purchase the updated maps and software. To discover more about Streets & Trips, go to www.microsoft.com/streets.

Garmin Mobile PC

In 2008, Garmin announced it was entering the laptop navigation-software market with a product called Mobile PC. The program brings the intuitive interface and look and feel of the company's automotive GPS products to Windows laptops. Versions of Mobile PC are available with or without a Garmin USB mouse GPS. To learn more about Garmin Mobile PC, visit www.garmin.com.

For information on other laptop navigation programs, check out the forums at www.laptopgpsworld.com and www.gpspassion.com/forumsen/default.asp?CAT_ID=4.

Chapter 14

On the Ground with TopoFusion

*I*f you spend a lot of time off the beaten path, say in the mountains, hills, deserts, or plains, you should consider using topographic mapping software. These programs use United States Geological Survey (USGS), or similar, digital maps, and are a perfect companion for any outdoor enthusiast. Before you head off into the wilds, you can print a topographic map of an area you're interested in visiting and even plan your trip on a PC. If you have a GPS receiver, you can interface it with the map program to see exactly where you've been. You can also upload waypoint and route information entered on the electronic map to your GPS unit.

Commercial topographic software packages are easy to use, convenient, and economical. Most of the map products on the market cost under $100 and provide maps for a single state, region, or beyond. Considering that a paper version of a USGS topographic map costs around $7, digital maps can be quite a bargain.

In this chapter, I discuss commercial map products that offer topographic maps of the United States. I primarily focus on a nifty program called TopoFusion. At the end of this chapter, I briefly review some other popular topographic map programs.

Discovering TopoFusion

Scott Morris was frustrated with commercial map software. As an avid mountain biker who frequently used a GPS receiver and digital topographic maps to plan routes and see where he had been, he thought he could create a better map program; one more suited for a serious outdoors user. So he and his brother Alan developed TopoFusion (it didn't hurt that Scott was also a Computer Science PhD candidate at the University of Arizona at the time).

TopoFusion is one of the more innovative, feature-laden, and easy-to-use top-ographic map programs. It's unique in that instead of coming bundled with CDs or a DVD filled with maps, it relies on a number of publicly available, free Internet data sources to display a wide variety of map types. You specify an area you want to see, and then TopoFusion downloads and shows the maps. Simple as that.

Obviously, you need an Internet connection to use TopoFusion, the faster the better. Also, if one of the map servers the program uses is down for some reason, you won't be able to access map data from that particular site. TopoFusion does cache previously viewed maps on your hard drive though, and they're available even when the Internet isn't.

Here are some of the things you can do with TopoFusion:

✔ View USGS raster topographic, Canadian vector topographic, Landsat (satellite), USGS aerial photo, and Tiger street maps

✔ Display three-dimensional maps based on elevation

✔ Blend different types of maps

✔ Create custom maps

✔ Import and export GPS waypoints, tracks, and routes

✔ Display geotagged digital photos on maps with a feature called PhotoFusion

✔ Track and analyze GPS, heart rate, and PowerTap (cycling) data for sports training

The two versions of TopoFusion are Basic and Pro. The Basic version uses older Microsoft DirectDraw technology, has a rudimentary set of features, and costs $40. TopoFusion Pro uses faster Direct3D graphics routines, has many advanced features, and costs $69.95. I recommend shelling out the extra bucks and getting the Pro version.

To learn more about TopoFusion, or to download a free evaluation version, visit www.topofusion.com. (The evaluation versions are fully functional and don't expire after a certain amount of time. The only hitch is the word DEMO appears on one-third of the displayed map tiles.)

Read on to take a look at some of TopoFusion's basic features and see how to use them. Whether you put on your boots and backpack (or saddle up your mountain bike) is completely up to you.

Displaying Maps and Finding Places

When you run TopoFusion for the first time, a map of the world displays (as shown in Figure 14-1). You can use various commands to zoom in on an area you're interested in.

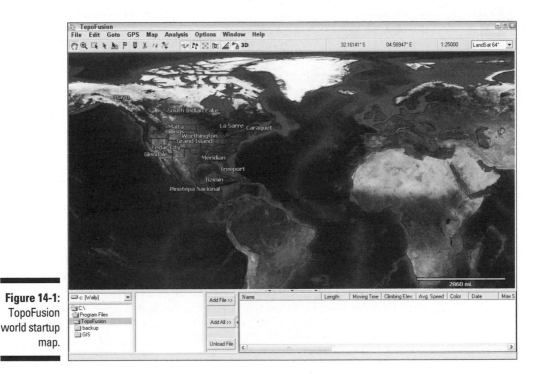

Figure 14-1:
TopoFusion
world startup
map.

Say you've got a trip planned to the Olympic National Park in Washington State. You're going to visit Hurricane Ridge, a place with some spectacular scenery and views, and do a little hiking.

Because you're the prepared type, you want to check out a topo map of the area before you go. You know to get to Hurricane Ridge you drive to the city of Port Angeles then take a steep road outside town that winds its way up into the mountains. Here's how you'd use TopoFusion to display a map.

1. From the Goto menu, choose Goto City.

A dialog box appears. (You can also choose Goto Coordinate from the Goto menu to displays a map based on coordinates of a known location.)

2. **Enter the name of a city near the place you want to go and click Search.**

   In this example, enter **Port Angeles**. When you click Search, a list of places that match your search text displays.

   If you enter a city and it can't be found, select the GIS Layers item in the Window menu. Then download the All US Cities file, which is a database of all U.S. cities and locations. By default, TopoFusion comes loaded with only locations of large U.S. cities.

3. **Select the city and click the Goto button.**

   The map centers on the location of the city.

Navigating a TopoFusion Map

One of the biggest advantages to using a digital map is you can interact more with it than you can with a paper map. For starters, here's how to change the map type, move around in a TopoFusion map, and change the map size.

Changing the map type

The very first time you use TopoFusion, it displays satellite imagery. However, that low-resolution satellite photo isn't going to be much use hiking, so you need to increase the detail. TopoFusion allows you to view a number of different other map types, including topographic maps, which is what we're interested in now.

In the upper-right corner above the map is a drop-down list containing available map types, including different scales. When you choose a map type, TopoFusion downloads and displays that map.

When TopoFusion downloads a map, it saves a copy on your hard drive in a cache file. The program always checks to see if map data for the area you want is in the cache, so you're not downloading the same data over and over. This saves time and bandwidth, and means you don't need an Internet connection to view maps you've previously displayed. You can specify how many megabytes (or gigabytes) of data to cache in the Preferences dialog box.

You have to use a little common sense in selecting a map type. For example, if you view a location in the United States and select Canadian topographic maps, the data isn't going to be too useful.

For example, start out by selecting Topo 64M (the M stands for thousand, as in a 1:64,000 scale map). Figure 14-2 shows what the map looks like. That's a better scale, with more detail shown.

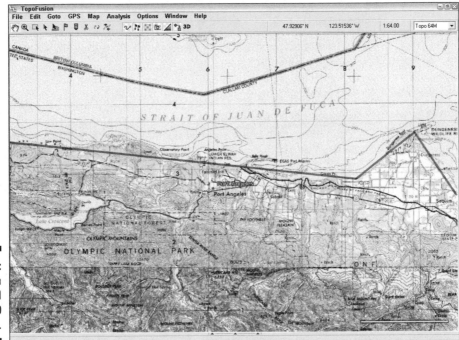

Figure 14-2:
TopoFusion
64M
(1:64,000
scale) map.

Moving around in a map

You can move around in a TopoFusion map two ways:

- ✔ **Click the Pan tool on the toolbar.** (It's shaped like a hand.) Move the cursor to the map and hold down the left mouse button; then scroll the map by dragging.
- ✔ **Use the arrow keys.** The right, left, up, and down keyboard arrows move the map in the associated direction.

While you move around, TopoFusion downloads map tiles for the area you're viewing on the screen.

While you move the cursor on the map, the latitude and longitude at the cursor are displayed in the toolbar above the map. This is useful for determining the exact locations of features on the map.

Changing the map size

TopoFusion offers a couple of ways for you to change the size of a map and show more detail or area.

- ✔ **Zooming:** The toolbar has an icon with a magnifying glass. Clicking the icon changes the cursor to a magnifying glass. Left mouse click to zoom in, right mouse click to zoom out. You can also hold the left mouse button down and drag to draw a rectangle around the area you want to zoom. If you have a scroll wheel on a mouse, you can also use it to zoom in and out.

- ✔ **Changing the scale:** In the drop-down list where you choose the type of map to be displayed, you can also change the scale.

For example, Topo 4M is a good size for viewing detail and planning. Figure 14-3 shows this level of detail around the Hurricane Ridge Visitor Center.

When you run TopoFusion, it remembers and displays the map you were last viewing just before you quit the program.

Planning a Trip with TopoFusion

If you've followed along to this point, you've located a map of Hurricane Ridge. Keep your imagination going and plan a hiking trip there. Suppose that your friends suggest you have to take a short hike up Hurricane Hill. They say it's past the visitor center, only a mile or two beyond the end of the road. However, the last time you listened to their directions, the short pleasant hike they described turned into an eight-hour death march through thick underbrush and straight up a rock face. This time, you decide to use TopoFusion to get a better picture of this particular outing.

1. **Look on the map for the end of the road and Hurricane Hill.**

   There it is. And, you see a trail that leads to the top.

   The symbol for a trail on USGS maps is a single dashed line. Lines with two sets of dashes indicate an unimproved road.

2. **Click the Mark Waypoint tool on the toolbar, move the cursor to the end of the road, and then click to create a waypoint for the trailhead (the beginning of the trail).**

   The Mark Waypoint tool looks like a single flag.

   A dialog box displays. Enter the waypoint name *trailhead* and then click OK.

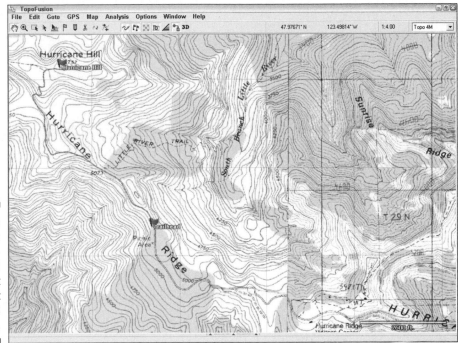

Figure 14-3:
A trail map with way-points and track that can be sent to your GPS receiver.

3. **Use the Mark Waypoint tool to create another GPS waypoint at the top of Hurricane Hill.**

   This marks a waypoint at the summit of Hurricane Hill; be sure to give the waypoint a meaningful name. Hurricane Hill would be a good, obvious choice.

   With these GPS waypoints set, you now know where the trail starts and ends. If you're using a GPS receiver, the first waypoint will help you find the trailhead, and the second waypoint will help you reach your final destination. (It's a popular trail and only a short distance, but since you used to be a Scout, you want to be prepared.)

   Be sure that the map datum matches the datum your GPS receiver is using (head to Chapter 4 for information about what happens when datums don't match). You can set the map datum in TopoFusion by choosing Options⇨Preferences⇨Units.

4. **Click the Draw Track tool on the toolbar to draw your planned course of travel on the map.**

The Draw Track tool, which looks like a pencil, works by drawing a line from the last place you clicked. However, it doesn't allow you to draw freehand like with a real pencil. Make sure Show Tracks is selected in the Map menu.

5. **Follow the trail by clicking the mouse (like playing connect-the-dots).**

   Trace the trail that heads up to Hurricane Hill, starting at the end of the road. After you click, the current length (in feet or miles) of the track displays in the status bar below the map.

6. **When you're finished, right-click and choose Save As from the pop-up menu. Name the file Hurricane Hill.**

   The finished map with the waypoints and the track is shown in Figure 14-4. You can upload the waypoints and track data to your GPS receiver before you leave on the hike to help with your navigation.

To edit a track, click on the Select Point tool, which looks like an arrow. A series of points displays on the track. You can delete and move individual track points.

Even if you don't have a GPS receiver to upload waypoint, route, or track data, you can still use TopoFusion to find a trail, print a map of that trail to take with you, and determine the distance of your hike. Just remember to bring your compass!

Understanding Terrain Elevation

Unless you've had some experience reading topographic maps, trying to figure out terrain elevation based on contour lines can be challenging. (Contour lines are a way of representing height on a map.) TopoFusion has two features — 3-D and a Profile tool — that help you better visualize where the land starts to get steep or flatten. You can use the Hurricane Hill map from earlier sections of this chapter to see just how much of a climb you're in for.

Displaying a 3-D map

Use TopoFusion to display 3-D, shaded relief images of a map to help you better understand the terrain. To show a 3-D image of the map that's currently displayed onscreen, click the 3-D button on the toolbar. (It's labeled 3D.)

The topographic map of Hurricane Hill (including the waypoint markers and track added in the earlier section, "Planning a Trip with TopoFusion") displays in 3-D, as shown in Figure 14-4.

Figure 14-4:
A Topo-
Fusion 3-D
map.

Here's how to control the view of the 3-D map with the Pan (hand) tool selected:

✔ Click and drag the cursor on the map to the right or left to rotate the image

✔ Click and drag the cursor up or down to change the vertical perspective

To toggle back to a 2-D version of the map, click the 3-D toolbar button.

Charting elevation profiles

Although the 3-D map view gives you a visual sense of how steep your hike is, you can get information that's even more detailed by using the Profile tool. This feature shows you the elevation gain/loss of your hike in pictures and numbers. Here's how to use this tool:

1. **Make sure that the map is displayed in 2-D.**

   See the preceding section for how to do this.

2. **Click the Profile tool on the toolbar (an arrow with a squiggly line beneath it; that's an elevation profile if you were wondering).**

3. **Select the track you want to profile.**

   A Profile dialog box opens, as shown in Figure 14-5, that displays an elevation chart of the trail as well as how many feet you'll be climbing and descending over the course of the trip.

If you've imported a track file recorded on your GPS receiver, each of the track points has elevation data associated with it, which TopoFusion uses to display the profile. If you create a track file in TopoFusion with the Draw Track tool and want to view the elevation profile, you first need to download elevation data associated with the map. In the Analysis menu, choose Climbing Analysis and click the Download DEM Date for Track button.

The Profile tool also displays elapsed time, speed, grade, and other information from a track file imported from a GPS unit. In addition to showing track data, there's an animated playback feature that simulates movement on the map based on recorded GPS track points.

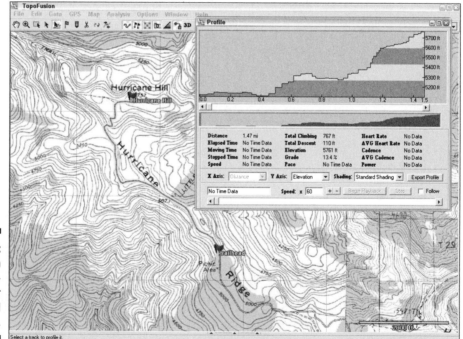

Figure 14-5:
Elevation profile information for a selected track.

Blending maps and aerial photos

TopoFusion has a unique feature that allows you to blend topographic maps with aerial imagery. This can be very useful because you can get a birds-eye photographic view of a location with overlaid contour lines, roads, features, and descriptive names. A blended map is shown in Figure 14-6.

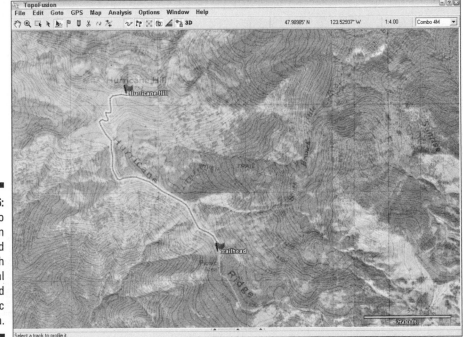

Figure 14-6:
A Topo
Fusion
blended
map with
aerial
photo and
topographic
data.

To create a blended map, choose a Combo type map from the drop-down list in the upper-right corner.

To change the map's appearance, choose Image Processing from the Window menu. In the Image Processing dialog box, you can specify the type of maps to blend and transparency, brightness, and contrast.

TopoFusion-blended maps can be displayed in 2-D or 3-D by toggling the 3D button.

I just scratch the surface on TopoFusion features; I suggest you view the user guide on the TopoFusion Web site or visit the user community forums at www.topofusion.com/forum. Scott Morris is constantly adding new features to TopoFusion and is very responsive to user suggestions for enhancements.

Reviewing Other Topographic Map Software

TopoFusion isn't the only commercial program available for working with digital topographic maps. DeLorme, Maptech, and National Geographic also offer software that's suited for recreational and trail use. Most topographic map programs have many of the same basic features; the biggest difference is their user interface. Because I don't have the time to cover all mapping programs in detail, this section provides a brief description of some popular programs that come bundled with maps, shows a few screenshots to give you an idea of what the user interface is like, and provides manufacturer Web sites to get more product information.

Some manufacturers offer demo versions of their programs so you can evaluate them. If you're interested in purchasing a program that doesn't have a demo available, be sure to check customer comments and reviews on Amazon (www.amazon.com) and Google (do a search for the product name and *review*).

DeLorme Topo USA

DeLorme (www.delorme.com), well known for its popular Street Atlas USA road navigation software, also offers a Windows topographic mapping program called Topo USA. DeLorme uses the same user interface (shown in Figure 14-7) for its entire line of consumer mapping products, so if you're a happy user of Street Atlas USA and are looking for a map program you can use off-road, it makes sense to keep things in the DeLorme family.

Topo USA comes on a DVD and contains topographic maps of the entire United States. The Topo USA maps are *vector-based* (drawn with lines and shapes) versus scanned paper maps. Topo USA can interface with GPS receivers, display 3-D maps, search for locations, provide elevation profiles, and plan for trips just like other map programs.

The latest versions of Topo USA work with DeLorme's handheld GPS receivers. That means any map you display on your PC can be transferred with all of the detail to a compatible DeLorme GPS unit. I discuss this more in Chapter 11.

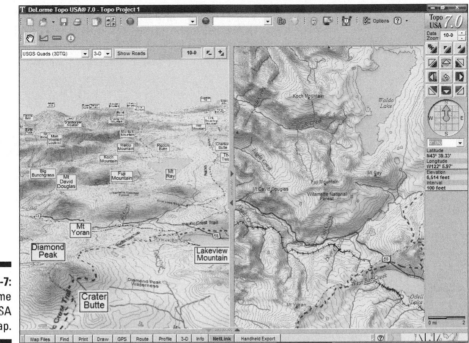

Figure 14-7:
A DeLorme
Topo USA
map.

Topo USA offers the following advanced features:

- ✔ **Street and POI data:** DeLorme includes updated street information and over 4 million POIs from its Street Atlas USA program.

- ✔ **Digital photo geotagging:** You can geotag digital photos with a GPS track file and embed thumbnail photos of locations on the map.

- ✔ **Aerial flyovers:** Get a birds-eye view of 3-D terrain with a simulated flight over a selected route.

- ✔ **Custom maps:** Topo USA comes with an extensive collection of drawing tools for editing maps. You can even add your own routable roads.

- ✔ **Online map purchasing:** Topo USA's NetLink feature allows you to purchase and download satellite imagery, USGS scanned maps, and NOAA scanned charts that can then be viewed from within the program.

Topo USA retails for around $100. Check out DeLorme's user forums at
`http://forum.delorme.com` for more product information.

Maptech Terrain Navigator

Maptech's Terrain Navigator (www.maptech.com) was one of the first commercial, Windows topographic map programs. (In the old days, it was known as *TopoScout.*) Over the years, Terrain Navigator has evolved into a sophisticated, powerful, electronic mapping tool with an easy to use interface.

Terrain Navigator uses scanned USGS maps and displays 2-D and 3-D views (a pair of 3-D glasses is even included), interfaces with GPS units to exchange waypoints, tracks and routes, and provides a number of options for printing maps. (An example Terrain Navigator map is shown in Figure 14-8.)

Versions of Terrain Navigator are available that provide topographic maps for individual states and regions of the U.S. (the program is the same, the map data bundled on CD-ROMs is different). Complete map coverage for a state, including 1:24,000 and 1:100,000 scale maps, costs around one hundred dollars.

Maptech has a free demo version of Terrain Navigator that comes with a single map of a wilderness area in Colorado. You can download the demo at www.maptech.com/support/downloads.cfm. Read the online help and tutorial to learn about all the program features.

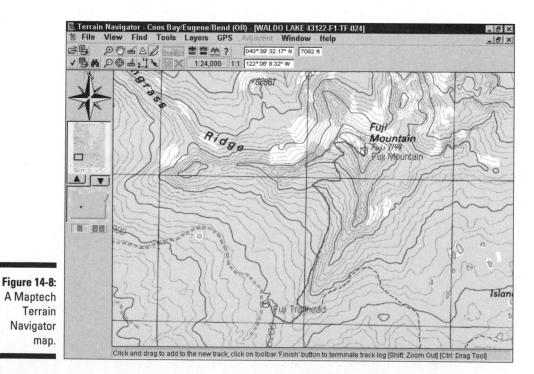

Figure 14-8:
A Maptech
Terrain
Navigator
map.

In addition to Terrain Navigator, Maptech has several other topographic map products that you might find useful, including

> ✔ **Terrain Navigator Pro:** Maptech's high-end version of Terrain Navigator targets professional map users and is priced around $300 per state. The program uses the same maps but supports aerial photo downloads, street address searches, map exports to GIS programs, and other advanced features.
>
> ✔ **Pocket Navigator:** A mapping program for Pocket PCs that allows you to load detailed Maptech digital topographic maps and nautical charts on your PDA, which you can then optionally interface with a GPS receiver for moving maps. The product is priced around $50.

National Geographic TOPO!

Many people got their first introduction to maps from *National Geographic* magazines. Each month, a carefully folded paper map of somewhere foreign and exotic came nestled inside the magazine. National Geographic is still in the map business, now offering several topographic map programs.

I was one of those kids hooked on National Geographic maps when I was growing up, and I still like them today. The company's map software is easy to use and strikes a nice balance of having just the right number and types of features, without having too many whistles and bells.

National Geographic software (`www.natgeomaps.com`) uses scanned paper maps and has all the same basic features as other mapping programs, such as GPS support, route planning, printing maps, and searching for locations. One bonus for Apple users is both PC and Mac versions of map software are available.

National Geographic's TOPO! State Series of software provides 1:24,000 scale and overview maps for individual states and multi-state regions. (A screenshot of a TOPO! 1:24,000 scale map is shown in Figure 14-9.) Each state or region retails for $99.95.

During the summer of 2008, National Geographic released a new program to the TOPO! product family called TOPO! Explorer. Instead of coming with a large, bundled collection of maps, the program features an *a la carte* approach, allowing users to purchase only the maps they need from a companion Web site (`www.topo.com`) — the software comes with credits for downloading 25 free maps. TOPO! Explorer is priced at $24.95. A Deluxe version is also available for $49.95 that includes 1:100,000 scale maps of the entire United States in addition to the 25 map download credits. Registered users of other TOPO! products can download a free version of TOPO! Explorer minus the map credits.

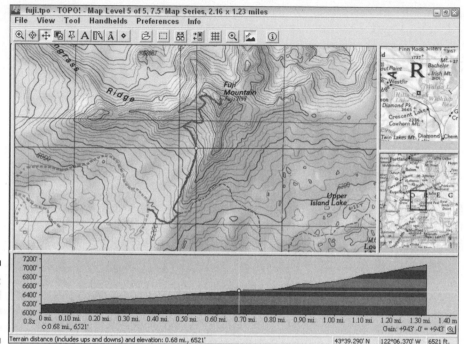

Figure 14-9:
A National
Geographic
TOPO! map.

The TOPO! Explorer Web site is an outdoors community site and a place to download maps for TOPO! products. Users can freely view online National Geographic maps and satellite imagery as well as access and post photos, trip reports, and GPS waypoints and tracks.

National Geographic also offers two other topographic map programs:

- ✔ **Weekend Explorer 3D** offers the same types of maps found in the TOPO! State Series, but only provides limited coverage of recreation areas located around selected major U.S. cities. Priced at $29.95 each, these programs are ideal if you plan to visit only popular outdoors destinations for short trips.

- ✔ **Back Roads Explorer 3D** contains 1:100,000 scale topographic maps and larger overview maps of the entire United States. In addition to terrain data, the software also has road and street data that you can overlay on a map, including paved and unpaved roads. Just be aware the level of detail on these maps is nowhere near other National Geographic products. The software retails for $49.95.

Chapter 15

From the Sky with Google Earth

*W*hen I was a kid, I always liked playing with desktop globes before geography class started. I'd find exotic sounding places mentioned in the news, compare the size of countries, or sometimes just spin the globe and stop it with my finger and see where I ended up. I could be Magellan and explore the entire world, all before the bell rang and Mr. Solberg began the serious business of education.

When Google released its Google Earth virtual globe program, it was like being a kid again, only better. Here was an easy-to-use program that combined with a speedy Internet connection to let me go exploring the world once again. Only this time, I could see detailed overhead images of cities, rivers, mountains, and even my own backyard. I could draw lines and measure distances. I could type in places like Tajikistan and the globe would spin by itself and take me there. It was like having access to my own spy satellite.

Google Earth is an amazing tool, and in addition to its entertainment and educational value, is pretty darn practical. Over the years, I've used it for overseas humanitarian work, disaster response, and trip planning. You're only limited by your imagination in how you can use this ultimate map program.

In this chapter, I introduce you to Google Earth's features, discuss some basic commands, and share tips and resources for getting the most from the program.

Discovering Google Earth

To begin, Google Earth is a virtual globe. It has color, satellite imagery of the entire planet. You can rotate the globe to a place you're interested in and then zoom in to view more detail. The program's interface is shown in Figure 15-1.

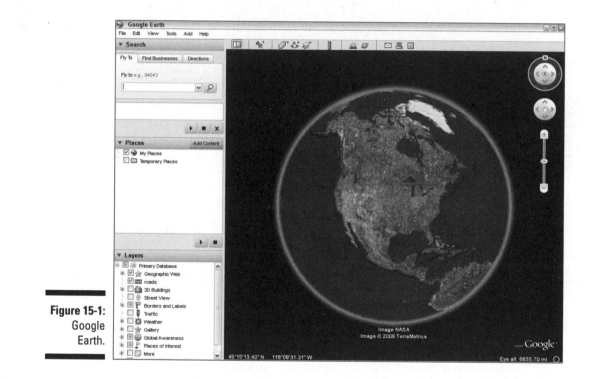

Figure 15-1: Google Earth.

Google Earth started life as a commercial software product called Earth Viewer. It was originally developed by Keyhole Inc. (a clever name for a business because Keyhole was the codename for a secret, U.S. government surveillance satellite). Google bought Keyhole in 2004, renamed their program, and began offering free and paid versions of Google Earth in 2005.

In addition to showing satellite imagery, Google Earth also supports overlaying data onto the overhead images. A wide variety of data layers are available including roads, current weather conditions, Bigfoot sighting locations, cities, borders, projected sea level increases, and the list goes on.

Between satellite images and layers, that's a staggeringly huge amount of data, which would be practically impossible to package and put in a box like a typical software product. To address this, Google keeps all the data online on servers. The program uses an Internet connection to download data you're interested in viewing. Rotate the globe a little to the left or zoom in, and Google Earth retrieves the data from a server and displays it.

Having access to these satellite images is incredible, but what is even more amazing is that the program is free. Versions are available that run on Windows, Mac, and Linux computers. To download a copy, as well as get more information, visit `http://earth.google.com`.

People are always finding new and creative ways to use the program and Google frequently makes new satellite images available and adds new features. One of the best places to keep up with Google Earth news is the Google Earth Blog at `http://gearthblog.com`.

Traveling the Globe

Google Earth is a little more complicated to use than your average school globe, so in this section, I step you through some of the basics of finding your way around Google's virtual world.

For starters, I need to dispel a popular misconception. The satellite images you see with Google Earth are not real time. If you hear your neighbor mowing the lawn, you can't rush over to your computer, fire up Google Earth, and see him. You need to work for one of those government three-letter agencies to have that kind of power at your disposal. The satellite images are snapshots taken over the past several years.

Now that I've cleared that up, here's how to use Google Earth.

Beyond free

In addition to the free version of Google Earth are two commercial variations with enhanced features:

✔ **Google Earth Plus:** Supports GPS receiver integration, higher-resolution printing, higher download speeds, and customer e-mail support. The added features cost $20 a year.

✔ **Google Earth Pro:** A business-oriented version of Plus that includes features such as movie making, GIS tools and data export, and enhanced printing. There is an annual $400 subscription fee.

You can learn more about these two versions at `http://earth.google.com`.

Keep in mind that if you just want to interface your GPS receiver to Google Earth to use it as a moving map, you don't necessarily need to pay for the Plus or Pro versions. Check out a free program called GooPS (`http://goops technologies.com`) that interfaces GPS units with the free version of Google Earth.

Navigation controls

To the right of the globe are three navigation controls, as shown in Figure 15-2. They allow you to move around, and zoom in and out on the earth.

Figure 15-2:
Google
Earth
navigation
controls.

Google changed the navigation controls in version 4.3 and these controls look and work a little different than the controls in older versions of Google Earth.

- ✔ The tool at the top (the circle with the eye in the center) orients your view.
- ✔ The second tool (the circle with hand in the center) controls movement.
- ✔ The bottom tool (the slider) controls zooming in and out.

I talk more about these controls next, but in the meantime, don't panic if while you use Google Earth, these controls dim and fade away. You're not seeing things and there's nothing wrong with your computer. This is the user interface allowing you to see more of the globe without it being obscured by controls.

To get the navigation controls back, simply move the cursor to the upper-right corner of the window.

Zooming in and out

The default Google Earth globe view is neat, in an astronaut kind of way, but you probably want to zoom in and look at things in a little more detail. Here's how.

To make the globe appear in a full window (as shown in Figure 15-3), deselect Sidebar from the View menu. To restore the sidebar, select it.

The simplest way to zoom in on a place is to move the cursor over the location and double-click. Each time you double-click, you get closer.

Figure 15-3:
Google
Earth
zoomed in.

In the bottom-right corner of the window is a number with the label Eye Alt. This is what the current view would look like if you were that many kilometers above the earth. (You can change the eye altitude to feet and miles in the Options dialog box if you're not fond of metric.)

When you start to zoom in, you see rectangular strips that are a little darker than the surrounding area. These strips indicate a higher-resolution satellite image is available. If you zoom in to an eye altitude of 1,000 kilometers, it's easy to tell at a glance which areas have high-resolution images.

Satellite image resolution in Google Earth ranges from around 15 meters to 15 centimeters (6 inches) for some metropolitan areas. Whenever you see satellite resolution numbers, they refer to the size a single pixel represents. Therefore, with 15-meter imagery, you could identify roughly a 45 x 45-foot object. With 15-centimeter resolution, you could see 6 x 6-inch objects on the ground. Generally, high-resolution is considered anything that's a meter or less.

If you want more control over zooming, use the slider navigation control.

- ✔ Click on the minus button to zoom out.
- ✔ Click on the plus button to zoom in.
- ✔ Drag the slider control between the two buttons (toward plus zooms in, toward minus zooms out).

Want to know when a satellite photo was taken? Starting with version 4.3, Google Earth displays the year (and sometimes the month and day) the satellite snapped the photo you're viewing. The date appears in the status bar at the bottom of the window after you zoom in a bit.

Moving around

You can move around in Google Earth several ways:

- ✓ **View orientation navigation control:** Click and drag the outer ring with N to change the direction north points. Click the arrows on the inner ring to rotate your view. This is like turning your head.

- ✓ **Movement navigation control:** Click an arrow on the circle with the hand in the center to move in that direction.

- ✓ **Keyboard arrows:** Up arrow scrolls up, right arrow scrolls right. You get the picture.

- ✓ **Mouse scroll wheel:** If your mouse has a scroll wheel, use it to zoom in and out.

- ✓ **Dragging:** Right click, hold the right mouse button down, and drag in the direction you want to move. Release the mouse button when you're done moving.

When you move the cursor around, the latitude and longitude of the cursor position appears in the status bar in the lower left of the window.

If you want to know what the shadows are like at a certain time of day, find the spot in Google Earth (versions 4.3 and higher), zoom in, and click the tool that looks like a sun coming over a mountain. A time scroll bar appears at the top of the window. Slide the control to a certain time of the day to see how much light there is. Flat locations with no surrounding terrain will uniformly get lighter and darker, but if there's a hill or mountain nearby that blocks the sun, you'll see its effects. Keep in mind Google Earth uses elevation data to determine shadows, and doesn't incorporate trees and buildings into its calculations.

Finding places

In addition to using the old-school, spin-the-globe method of exploring and locating places, Google Earth also offers a more high-tech approach to finding locations around the world.

To the top, left of the globe is a panel labeled Search with a tab labeled Fly To. You can search for locations several different ways:

Please fasten your seatbelts

Since version 4.2, Google Earth has featured a cool option that turns your PC into a flight simulator. You have the choice of flying an F-16 fighter or a high-performance Cirrus SR-22 over any of the satellite image's photo-realistic terrain. Here's an example of what you might see.

All the controls and basic flight instructions are at `http://earth.google.com/intl/en/userguide/v4/ug_flightsim.html`. To enter flight simulator mode, press the Ctrl+Alt+A keyboard combination. In version 4.3, the flight simulator was added to the Tools menu, making it easier to turn on.

- ✔ **City, State** such as Seattle, WA
- ✔ **City Country** such as Paris France
- ✔ **Number Street City State** such as 1600 Pennsylvania Ave Washington DC
- ✔ **ZIP code or Postal code** such as 97701
- ✔ **Latitude, Longitude in decimal format** such as 37.7, –122.2
- ✔ **Latitude, Longitude in Degrees, Minutes, Seconds format** such as 37 25'19.07"N, 122 05'06.24"W or 37 25 19.07 N, 122 05 06.24 W

Enter your search criteria and click the magnifying glass button. Google Earth will rotate the globe, fly you to that location, and zoom in (which can sometimes make you a little dizzy).

If there are multiple locations with the same name (say you entered a city without the state, and a number of U.S. cities have that name), Google Earth lists them all. Click the one you're interested in viewing.

There are also tabs for finding businesses and points of interest and for providing driving directions like you get from Web-hosted street maps, such as Google Maps.

If you ever encounter fuzzy images when you zoom in, it might not be your computer monitor. Google has blurred out high-resolution images of sensitive areas at the requests of some governments.

Getting a 3-D View

Getting a birds-eye view of a place is pretty slick, but birds don't just see in two dimensions, and neither does Google Earth. The program can display three-dimensional representations of the terrain, based on elevation data. An example 3-D view is shown in Figure 15-4.

To get a three-dimensional view of a place, press the Shift key and hold down the left mouse button. A crosshairs circle appears at the cursor location.

Now you have several options:

- ✔ Drag the mouse toward the bottom of the screen to tilt the view down and get a 3-D perspective.
- ✔ Drag the mouse toward the top of the screen to flatten the view so it appears you're directly overhead, looking down.
- ✔ Drag the mouse to the right to rotate the view to the right.
- ✔ Drag the mouse to the left to rotate the view to the left.

You can use a combination of these commands. For example, dragging the mouse down and to the right increases the tilt while rotating the view right.

Press U on the keyboard to restore the map to a directly overhead, looking down view.

Figure 15-4:
A Google
Earth 3-D
view.

If you can't get a 3-D view, make sure Terrain is checked in the Layers panel to the left of the globe window.

And that tip is a good segue into telling you all about layers.

Using Layers

Google Earth doesn't just display pretty pictures of the planet. The program can also overlay many different types of data, providing rich, interactive information for places all over the world.

It's worth mentioning that Google Earth doesn't limit itself to only showing you our third planet from the sun. If you click the toolbar button that looks like the planet Saturn, you start Google Sky, a full-blown astronomy program that lets you explore the universe and boldly go where no man, or woman, has before. There's talk of perhaps Google Ocean in the future, which will let you play Jacques Cousteau.

Data layers typically have icons, labels, or images that are drawn on the globe. In many cases, you can click on an icon to get information associated with it. Here are two examples:

✔ **If the Panoramio layer is turned on,** light blue dots appear on the globe. Click a light blue dot to see a digital photo someone took of that location. (`www.panoramio.com` is a photo-sharing site where people post digital photos of places all over the world with geographic coordinates.)

✔ **If the Wikipedia layer is turned on,** light purple dots appear on the globe. Click a light purple dot to see Wikipedia information about that place. Many cities have Wiki data; an example is shown in Figure 15-5.

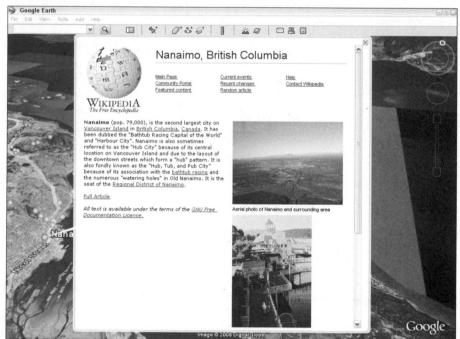

Figure 15-5: Information about a city from the Wiki layer.

Choose the data layers you want to see, and while you zoom in, associated icons start to appear.

Some layers have sublayers. For example, in the Places of Interest layer, you could decide to show only bank locations and not display coffee shops.

Available layers

Here's a list of layers that are available in Google Earth:

- **Geographic Web:** This layer displays data from the Google Earth Community, Wikipedia, Panoramio, and provides previews for data from several other layers.

- **Roads:** Google has road data for selected countries throughout the world. This layer displays streets and roads with their names.

- **3D Buildings:** Google has collected building dimensions for selected metropolitan areas. Checking this layer will draw buildings in 3-D, either photorealistic or shaded in gray. Zoom in to Seattle for an example.

- **Street View:** Google has been compiling street-level photographs of metropolitan areas for its Web-based, Google Maps. These photos are now available in Google Earth.

- **Borders and Labels:** This layer shows political boundaries, such as international borders, county names, and populated places.

- **Traffic:** With this layer, Google Earth provides real-time traffic information for major metropolitan areas. Zoom in on a freeway and green, yellow, and red dots appear representing normal, below normal, and slow traffic speeds. Click a dot to get details.

- **Weather:** You can check the weather in different parts of the world with this layer. Sublayers include Clouds, Radar, and Conditions and Forecasts.

- **Gallery:** Gallery contains photos and videos from a number of organizations and businesses, including volcano locations, YouTube videos, travel and tourism spots, and National Geographic content.

- **Global Awareness:** This layer contains data from organizations that deal with environmental and humanitarian issues.

- **Places of Interest:** Just like you can with an automotive GPS receiver, click a sublayer to display icons of restaurants, hospitals, stores, and other POIs.

- **More:** This is a catch-all layer for types of data that don't fit elsewhere, including U.S. government data, foreign language alternate place names, and available satellite coverage maps from Spot and Digital Globe (Google's primary imagery suppliers).

- **Terrain:** This layer contains terrain elevation data and allows Google Earth to create 3-D views.

Google often updates layer data and adds new layers, so don't be surprised if new selections show up in the Layers panel.

Turning layers on and off

Now that you have a better idea of what types of data layers are available, here's how to turn layers on and off.

In the Layers panel, each layer has a box to the left of it (refer to Figure 15-1 to see what I mean). If the box is

- **Blank:** The layer isn't displayed.
- **Checked:** The layer is displayed.
- **Filled:** The layer has sublayers and some are being displayed.

  Click the box to change the display status. For example, if you click a checked box, it becomes blank and the layer is no longer shown.

If a layer has a plus mark to the left of its name, different types of associated data can be displayed. For example, with the Borders and Labels layer, a number of sublayers are available that show different types of borders and labels.

- Click the plus button to expand the list of sublayers.
- Click the minus button to collapse the list and hide the sublayers.

 Google Earth has a number of layers turned on by default, and in some locations, the globe can become cluttered with all the layer icons. One of the first things I suggest doing is identifying the layers you plan to use and turning all the other layers off. You can always turn them back on again. Keep in mind that the more layers you have turned on, the more Google Earth will slow down from the additional downloading.

Adding Placemarks

Say you find some place cool in Google Earth and you want to mark the location. It's pretty easy to do; here are the steps

1. **Zoom in to the place you want to mark; the closer the better.**
2. **In the toolbar, click the Add Placemark button (it looks like a pushpin).**

   A pin appears on the map and an information dialog box displays.
3. **Click the center of the pin, hold the left mouse button down, and drag the pin to the place you want to mark.**
4. **Enter a name for the place and a short description.**

You can also change the placemark color and size, and other display information.

5. **Click OK.**

The placemark now appears on the globe (an example is shown in Figure 15-6). You can edit or delete a placemark by right-clicking it and choosing a command from the pop-up menu.

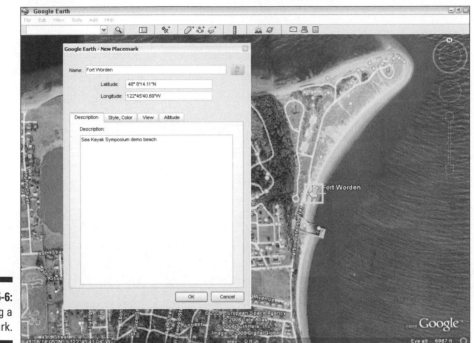

Figure 15-6:
Creating a
placemark.

Placemarks also appear in the Places panel to the left of the globe window. Just like layers, when a placemark is checked, it's displayed. When unchecked, it's hidden.

A large online Google Earth user community creates placemarks for newsworthy places and discusses new and interesting things found in the satellite imagery. Visit the forums at http://bbs.keyhole.com/entrance. php?Cat=0.

Whenever you see KML or KMZ mentioned on a Web site or forum, they refer to Google Earth file formats for storing placemarks, GPS data, and other information. KML stands for Keyhole Markup Language. KMZ is a KML file that has been compressed using the ZIP format. To learn more about KML go to http://code.google.com/apis/kml/documentation.

If you have GPS data in GPX format, you can easily import it into Google Earth and view waypoints and tracks. In the File menu, select Open then choose your data file.

Measuring Distances

Google Earth offers a couple of methods you can use to measure distances.

The first method is good for getting a rough sense of distance: From the View menu, choose Scale Legend to display a map scale at the bottom of the globe window.

A more accurate way to measure distance between two points is to use the Ruler tool.

1. **In the toolbar, click on the Ruler button.** (You can't miss it, it looks like a ruler.)

   The Ruler dialog box appears. You have two choices:

   • *Line* measures distance between two points.

   • *Path* measures distance between multiple points.

2. **Move the cursor to the starting point of where you want to start measuring and left click.**

3. **Move the cursor to the next point and left click again.**

   Google Earth displays the distance in the Ruler dialog box. (An example is shown in Figure 15-7.) You can use the Clear button to erase the distance or change the measurement type.

The Smoots measurement type is a Google in-joke. A *smoot* is a nonstandard unit of measure named after Oliver Smoot, who was part of an MIT fraternity prank in the 1950s. One smoot is equal to 5 feet, 7 inches, which happened to be Smoot's height at the time.

Google Earth offers much more than I have time to cover in this chapter. To learn about all of its features and commands, spend some time with the online user guide at `http://earth.google.com/intl/en/userguide/v4`.

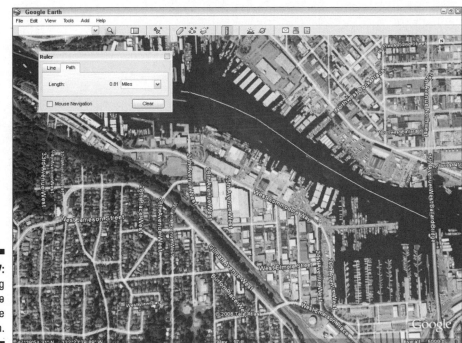

Figure 15-7:
Measuring
distance
in Google
Earth.

Microsoft Virtual Earth

Google isn't the only game in town when it comes to high-resolution satellite imagery. Microsoft also has a project called Virtual Earth that shares a number of features available in Google (and a few that aren't, such as birds-eye views taken by aircraft). Virtual Earth is integrated into Microsoft's Live Search Web site (`http://maps.live.com`), so you don't need a standalone program to view the imagery. Check out these two blogs for more information on Virtual Earth: `http://blogs.msdn.com/virtualearth` and `http://virtualearth.spaces.live.com`.

Chapter 16

Creating and Using Digital Maps with OziExplorer

OziExplorer is a popular, Windows shareware program that you can use to create digital maps. OziExplorer is widely used by recreationists, land managers, and public safety agencies. The program doesn't come bundled with a set of maps per se; instead, you can use it to view maps that are commercially or freely available on the Internet in a number of different data formats. You can also use Ozi to create customized, do-it-yourself, digital maps. In this chapter, I briefly describe some of OziExplorer's features, and then I focus on creating digital maps from scanned paper maps.

Discovering OziExplorer Features

OziExplorer is a powerful and versatile mapping program developed by Des Newman. (Newman hails from Australia, and *Ozi* is slang for *Australia* — get it?) Newman originally wrote the program for personal use during four-wheel-drive trips in the Australian outback. He released OziExplorer as shareware, which has evolved into a sophisticated mapping tool that's constantly updated. Some of the program's key features are that it

✔ **Interfaces with GPS receivers:** OziExplorer can communicate with just about any GPS receiver on the market that can connect to a PC, allowing you to upload and download waypoints, routes, and tracks to and from GPS units and PCs. (For more on waypoints, routes, and tracks, see Chapter 4.)

✔ **Works in many languages:** Localized versions of OziExplorer are available in a number of different languages, including English, German, French, Spanish, and Italian.

✔ **Provides real-time tracking:** If you have a laptop connected to a GPS receiver, OziExplorer displays a *moving map* with your real-time, current position and other travel information. (For more on using a laptop in tandem with a GPS receiver, see Chapter 10.)

✔ **Is easy to use:** OziExplorer boasts a large number of features, such as annotating maps and extensive import and export capabilities, all of which are easy to use. Figure 16-1 shows a demo map that comes with OziExplorer and the program's toolbar and menu-based user interface.

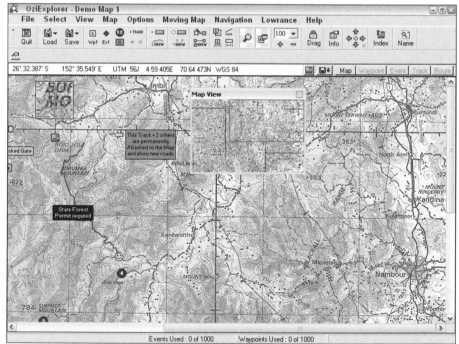

Figure 16-1:
A demo map in OziExplorer.

✔ **Supports an extensive number of map formats:** OziExplorer can access many popular digital map data formats (such as *DRG,* a Digital Raster Graphics topographic map) and can associate georeferenced data with common graphics file types. For example, you can take a file that you created in Paint or some other graphic program and turn it into a *smart map* (a map where geographic coordinates are associated with individual pixels). Loading graphics files and calibrating the data to create digital maps are discussed in upcoming sections of this chapter.

The best way to find out about OziExplorer's features is to download the program and try it. (OziExplorer works with PCs capable of running Windows 95 through Vista.) You can find it at www.oziexplorer.com. Two versions of the program come in the install package:

- ✔ **Trial:** The trial version is a limited version of OziExplorer that has all the program features enabled, except the program can't

  - Communicate with GPS receivers

  - Save or load waypoints, routes, or tracks

  - Save maps as image files

  - Run for more than an hour at a time

- ✔ **Shareware:** The shareware version has a number of features disabled and has the following limitations:

  - Only bitmap (BMP) images can be imported as maps (as opposed to many graphics file types in the registered version).

  - Only two points can be used to calibrate a map, reducing the potential accuracy. (Up to nine points are available in the registered version.)

  - Limited support is available for map projections, grid systems, and datums (compared with extensive support in the full version).

Between these two programs available in the install package, you can get a feel for OziExplorer's capabilities. And when you register the shareware version, the full monty of features is enabled. The $95 registration fee is a nominal investment considering the many features that the full version of OziExplorer offers.

If you're looking for a street and road navigation program, OziExplorer isn't the best choice. Ozi is more suited for adventures off the beaten path. If you need a program that helps you find the best route between two addresses on streets and highways, you're much better off using some of the commercial mapping programs that I describe in Chapter 13.

For more practical information on OziExplorer, including troubleshooting tips, check out the popular Yahoo! Groups e-mail list and forum devoted to the program at groups.yahoo.com/group/OziUsers-L.

In addition to OziExplorer, Des Newman also has two other related mapping programs:

- ✔ **OziExplorer3D:** This program is used in conjunction with OziExplorer to display maps in three dimensions. OziExplorer writes elevation data to a file, which OziExplorer3D uses to create a 3-D representation of the map. OziExplorer3D is priced at $30.

✓ **OziExplorerCE:** OziExplorerCE is moving-map software for Pocket PC/ Windows Mobile Edition PDAs. You create maps and plan trips with OziExplorer on your PC and then download the map data to your PDA to use with OziExplorerCE. When you connect your PDA to a GPS receiver, the program retrieves GPS data and displays your current location on a map. The PDA version of Ozi costs $30.

To discover more about the features of these two programs and download demonstration versions, go to www.oziexplorer.com.

Moving from Paper to Digital Maps

OziExplorer supports a number of different digital map types that use *geo-referenced data* (information that allows a program to precisely identify locations and coordinates on a map). However, one of the program's most powerful features is its ability to turn your graphics files into georeferenced maps. This means if you have a paper map, you can scan it, load it into OziExplorer, and effectively make it an electronic (digital) map. This is a three-step process:

1. **Scan the map.**

   Use a scanner to create a digital image of the paper map. Stitch individual map pieces, if necessary.

2. **Edit the map.**

   Make changes to the scanned map before it's used.

3. **Calibrate the map.**

   Load the edited map into OziExplorer and associate georeferenced data with the map image.

The following sections explore the above three steps in further detail.

Scanning and calibrating your maps can be fairly time consuming and some-times frustrating if you can't seem to get the map coordinates to match up with reality. Some maps are definitely easier to calibrate than others. If you're not technically inclined or are somewhat impatient, you'll probably want to stick to importing maps that are already georeferenced, such as freely available United States Geological Survey (USGS) DRG topographic maps. (See Chapter 20 for places on the Internet you can download free maps.)

Step 1: Scan the map

The first step when converting from paper to digital is to scan the paper map and turn it into a graphics file. You don't need an expensive, high-end scanner to accomplish this task; most any color scanner will work.

If you want to scan a large map — say, anything bigger than a legal size piece of paper — consider literally cutting it into pieces that will fit on your scanner (8.5 x 11 inches works well). Instead of using a pair of scissors, use a paper cutter, such as those found at copy centers, to ensure that you end up with straight cuts. The straight cuts are important for accurately aligning the map on your scanner. Although you can scan a large map one portion at a time without cutting it, it's more of a challenge to get the edges lined up when you stitch them together as I discuss next.

Here are some tips to improve your map scanning:

- **Use medium dpi:** Scanning the map between 125–200 dots per inch (dpi) is good enough; you don't need to scan at higher resolutions typically used for reproducing photos.

- **Use color photo scanning:** Most scanning software has different settings for different types of documents you want to scan, such as text, line drawings, and photographs. Select the color photograph option to retain the most detail. However, remember that most maps don't have millions of colors like photographs, so if your scanning software supports it, use a 256-color setting.

- **Watch edge alignment:** Place the to-be-scanned map directly on the scanner, ensuring that the edges are aligned directly against the scanner bed with no gaps. You need to keep the paper map as square as possible to reduce distortion during a scan. No sheet feeding, please.

- **Prevent edge distortion:** To help keep the map edges pressed flat, leave the scanner cover open and use a book or something heavy to set on top of the map. Typically, the edges are where the most distortion occurs during scanning because they tend to lift.

- **Experiment with settings:** Try a couple of experimental scans first, changing the brightness and contrast settings. If you're scanning a number of maps over a period of time, write down the settings that give you the best output so you can use them next time.

- **Save the final scan as BMP:** When you're ready to produce a final scan of the map, initially save it as bitmap (BMP) format file. This produces an image that's as close to the original map as possible; bitmap files aren't compressed like JPG and other graphics file formats. Bitmap files do take up a lot of memory and disk space, but after you edit a file, you can save it as a smaller graphics format.

If you have a map that's made of multiple image files, such as a large map cut into a series of smaller maps, you need to *stitch* them and save a single, large image. You can do this one of three ways:

- Commercial graphics programs, such as Adobe Photoshop and Corel Paint Shop Pro, have commands for combining files.

- Use a stitching program, such as AutoStitch (`www.cs.ubc. ca/~mbrown/autostitch/autostitch.html`), Reegemy (`http:// regima.dpi.inpe.br`), or StitchMaps (`www.stitchmaps.com`) to make the process much easier.

- Manually stitch together images with Microsoft Paint (or similar) by using the Paste From command on the Edit menu. Here's a link to a great tutorial on stitching together scanned images: `www.sibleyfineart. com/index.htm?tutorial--join-scans.htm`.

Step 2: Edit the map

After you successfully scan the map, make any last-minute changes to the image. This could include

- Adjusting the brightness and contrast to make the map more readable

- Adding symbols or text information

- Removing the white space (or *collar* as it's known in map-speak) that surrounds the map

Use your favorite graphics program to make any final edits to the map image. After you're through, save the map as a TIFF, PNG, or JPG file to reduce how much disk and memory space the image takes. (These compressed file formats are more space-efficient and memory-efficient.)

The shareware/demo version of OziExplorer can load only BMP images. Because bitmaps aren't compressed, the entire file must be loaded into memory, which can slow the performance of computers that don't have much RAM.

Step 3: Calibrate the map

After you scan, edit, and save your map, one more step is left before you can start using the map with OziExplorer. At this point, your map is simply a graphics file. You can use Microsoft Paint or any another graphics program to view, edit, and print the map, but you want to turn the image into a *smart map* to take advantage of OziExplorer's features.

OziExplorer and World War I

The 1980s movie *Gallipoli* recounted the Australian experience of fighting the Turks during World War I. Although long before the time of computers and mapping software, WWI also has a link to OziExplorer.

Howard Anderson wrote a fascinating article on using OziExplorer to locate old World War I trench lines in France and Belgium. The remains of the trenches are long gone, but by using old maps from the period, scanning them, and adding georeferenced data, Anderson was able to clearly determine where the trenches were during the early 1900s.

After he had scanned and georeferenced the old military maps, Anderson used OziExplorer to draw GPS tracks on personally created digital maps to trace the outlines of the trenches. He also used waypoints to identify military and land features. Anderson then took the tracks and waypoints and overlaid them on a modern map in OziExplorer. This revealed where the long-ago war emplacements once stood. Anderson's last step was to visit France with a GPS receiver and his old and modern maps. He found that, with relative accuracy, he could stand on the site of a trench where his grandfather had fought more than 90 years ago.

To read Anderson's complete account, which has historical insights as well as his experiences in using OziExplorer, visit the Western Front Association's Web site at `www.westernfrontassociation.com/thegreatwar/articles/trench-maps/stand.htm`.

Creating a smart map requires *calibrating* the map (also known as *registering* or *georeferencing*). This involves linking location data with the map image so that each pixel in the map has a geographic coordinate associated with it. When a map has georeferenced data, you can

✔ Move the cursor on the map, and OziExplorer will accurately report the coordinates of the cursor in latitude and longitude or UTM (Universal Transverse Mercator; read more about these in Chapter 2).

✔ Draw lines on the map to measure distance.

✔ Calculate the size of geometric areas.

✔ Track and display your current position on the map when the computer is connected to a GPS receiver.

✔ Transfer GPS waypoints, routes, and tracks between the map program and a GPS receiver.

When you calibrate a map with OziExplorer, the georeferenced data isn't embedded directly inside the image file. OziExplorer creates a separate MAP file (`.map`) that contains the following information:

✔ The location of the map image file

✔ The map datum

 The map projection

 Map calibration data

A MAP file is in text format and can be viewed with any word processor.

Calibration requires you to identify a series of points on the map with known coordinates. Depending on the number of points that you select, as well as the map datum and projection, OziExplorer performs different mathematical calculations to link coordinate information with the map image.

Some digital maps have georeferenced data embedded directly into the map image as tags or come with associated files that contain the reference data. (A common example is a DRG map.) OziExplorer can use these maps without going through the calibration process that I describe next. Check the OziExplorer Web site or the program's online help for a full list of these supported map types.

Choosing calibration points

To associate georeferenced data with your scanned map, you need to find a series of points on the map with coordinates you know. If you can't assign a latitude and longitude (or UTM) position to a few features on the map, you won't be able to calibrate the map.

Here are some ways in which you can pick calibration points:

- If the scanned map has a collar with coordinates printed on the edge, use the coordinate marks. USGS topographic maps are easy to calibrate because each corner is marked with the latitude and longitude.

- Use Google Earth (described in Chapter 15) to get the coordinates of a feature on a map that you want to calibrate. This can be a man-made feature (such as a building or a bridge) or a natural feature (such as a mountain peak).

- If you're in the United States, you can use National Geodetic Survey datasheets to identify points on the ground that have known coordinates associated with them. Visit www.ngs.noaa.gov/cgi-bin/datasheet.prl to access datasheets for your area.

- Visit a location that's clearly identifiable on the map, and use your GPS receiver to record the coordinates for that location. Road intersections make good calibration points. (Just watch out for traffic!)

The number of points that you select for calibration depends on the map and how much accuracy you want. Use the following guidelines to determine how many points you should use:

- **Two points:** If you're limited to two calibration points, select two points at opposite corners of the map. (This is the only calibration method available in the shareware/demo version of OziExplorer.)

If you have a registered version of OziExplorer, always select at least three calibration points, such as three corners of a map.

✔ **Three or four points:** Selecting three or four points (such as the map corners) provides better map accuracy and should be all you need for calibrating most maps.

✔ **Five or more points:** If the latitude and longitude lines are curved, if the paper map has been folded, or if the scan of the map is distorted, use more than four points. OziExplorer supports up to nine calibration points; generally, the more points you select, the better the accuracy with any map that might have distortions that could impact accuracy.

Try to spread your calibration points out over as much of the map as possible. If you clump the points together in a small area, the accuracy won't be as precise.

After you look at the map and determine which calibration points you're going to use, write down the coordinates and then double-check that they're correct. Entering calibration points with incorrect coordinates is a common cause of map accuracy troubles.

Figure 16-2 shows a 1:100,000 scale USGS topographic map. The map was scanned in pieces, stitched, and then saved as a JPG file. The original paper map had the latitude and longitude coordinates in all four corners, so I can use these coordinates as the calibration points. (This map is too large to reproduce here, so you can see the coordinates only in the upper-left corner.)

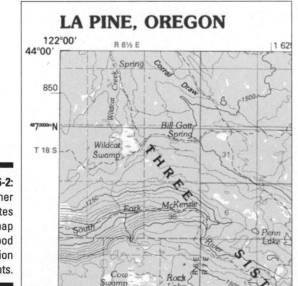

Figure 16-2: Corner coordinates on a map make good calibration points.

Setting calibration points

After you decide which points you'll use to calibrate the map, you load the map image file into OziExplorer and do some georeferencing. Here are the steps to take:

1. **Choose Load and Calibrate Map Image from the File menu; a file dialog box displays.**

2. **Select the location of the map image file and click Open.**

   The map appears in a setup window, as shown in Figure 16-3.

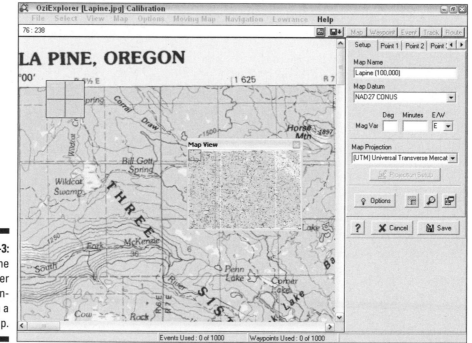

Figure 16-3: The OziExplorer setup window with a loaded map.

3. **Enter the map's name, datum, and projection.**

   Information provided on the example paper map says the datum is NAD 27 and the projection is Universal Transverse Mercator. Enter those values. (To read about map datums and projections, check out Chapter 2.)

   If you don't know the map datum, make an educated guess or use WGS 84. If you don't know the map projection, try using Latitude/Longitude. Both of these settings can be changed later if they end up incorrect.

Entering the wrong map datum and projection can severely affect the accuracy of your map. If you don't know the datum and projection, your best bet when calibrating the map is to match coordinates that you recorded with your GPS receiver to features that you can clearly identify on the map; then use the WGS 84 datum and Latitude/Longitude projection.

4. **For each calibration point (up to nine points), repeat the following steps:**

    a. *Click the point's tab in the setup window.*

    Figure 16-4 shows Point 1.

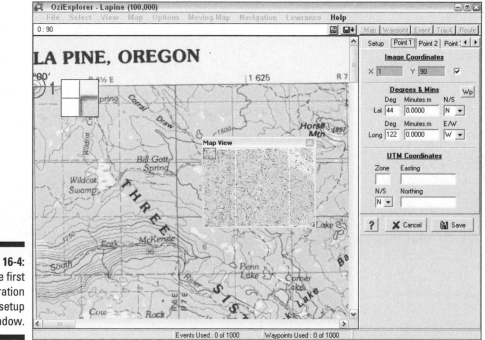

Figure 16-4: The first calibration point setup window.

    b. *Move the cursor on the map over the point.*

    The cursor turns into a cross-hair icon with the number 1 next to it. A zoom window shows a magnified image of the cursor location allowing you to place the cursor precisely over the calibration point.

    c. *Click the left mouse button to select the calibration point.*

    A bull's-eye appears over the calibration point.

    After the bull's-eye appears onscreen, you can move the calibration point to a new location by holding down the Shift key and using the navigational arrow keys. This gives you a fine level of control over placing the calibration point.

   d. *Enter either the latitude and longitude or the UTM coordinates for the calibration point.*

   OziExplorer expects the latitude and longitude coordinates to be expressed in degrees and decimal minutes. If your coordinates are in another latitude and longitude format, check out some conversion utilities I describe in Chapter 12.

5. **When you're finished, click Save.**

   OziExplorer prompts you for a name and where to save the MAP file (which contains the georeferenced data). You're now ready to start using your imported map. The example scanned-and-calibrated map being used with tracks and waypoints is show in Figure 16-5.

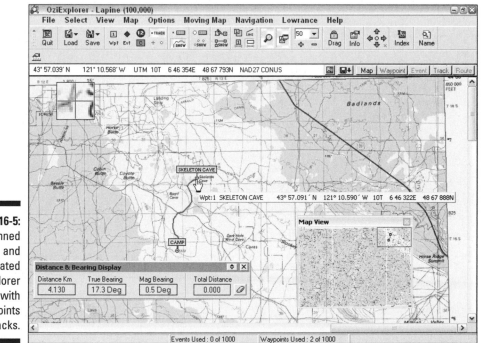

Figure 16-5: A scanned and calibrated OziExplorer map with waypoints and tracks.

Maps that are made for general information purposes (such as park visitor maps or maps in textbooks) are usually difficult to calibrate correctly because they don't have the high level of accuracy as navigation or survey maps do. (In fact, they're more of diagrams — not maps — because they don't have projections.) The actual locations of features shown on such maps might not be correct, and distances and proportions might not be accurate.

Checking your work

After you load and calibrate a scanned map, always check to make sure that your calibration was correct. Here are some ways to do this:

✔ Choose Grid Line Setup from the Map menu. This allows you to overlay latitude and longitude grid lines over the map. The overlaid grid lines should match with the grid lines on the map or the coordinate marks on the map edges.

On USGS topographic maps, don't get latitude and longitude lines confused with the more predominant Township and Range lines. (To find out the difference between latitude and longitude and Township and Range, read Chapter 2.)

✔ If your scanned map has a ruler printed on the map that shows distance (for example, in 1-mile increments), choose Distance Display from the View menu. This allows you to measure the distance of a line that you draw on the map. The length of the map's ruler should match whatever distance is shown when you measure it.

✔ Move the cursor over a feature on the map with known coordinates. (The coordinates can come from a GPS receiver, another mapping program, or a gazetteer.) The cursor coordinates in OziExplorer should be relatively close to the known coordinates of the feature.

If any coordinate or distance numbers seem significantly off, there's likely a problem with your calibration. To address this, do the following:

1. **Choose Check Calibration of Map from the File menu.**

   This displays the calibration setup window.

2. **In the calibration setup window, verify or change calibration data.**

   • Make sure that your calibration points are in the proper location on the map and have the correct coordinates.

   • Try changing the map datum and/or projection to different values.

3. **Click Save to save the results to a different MAP file.**

   The newly calibrated map is loaded and displays onscreen.

4. **Check whether the calibration is more accurate.**

   If the calibration isn't sufficiently accurate, repeat these steps.

FUGAWI

FUGAWI Global Navigator is a popular Windows mapping program with some of the same basic features as OziExplorer.

Like OziExplorer, FUGAWI can read a number of different map formats with georeferenced data as well as use do-it-yourself, calibrated, scanned image maps. The program can also interface with a GPS receiver and handle waypoints, routes, and tracks. See a FUGAWI map in the figure here.

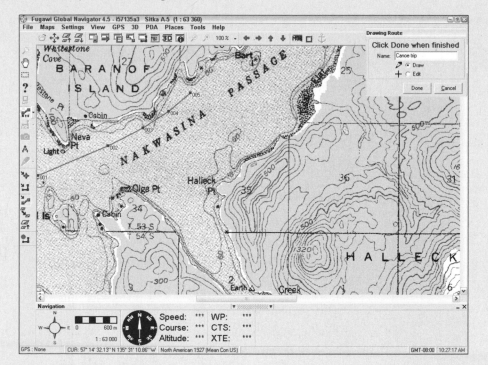

Comparing FUGAWI with OziExplorer, you'll see that FUGAWI isn't shareware, it's priced a little higher, and it comes bundled with a base set of maps.

FUGAWI and OziExplorer both have extremely loyal user bases. If you check around the Internet, you'll see some strong feelings expressed about why one product is better than the other. (I think OziExplorer has a better user interface and supports a few more advanced features.) If you're considering FUGAWI or OziExplorer, download the free demo versions of each program and see which one you prefer.

For more information about FUGAWI and to download the demo version of the software, visit www.fugawi.com.

Part IV
Using Web-Hosted Mapping Services

The 5th Wave   By Rich Tennant

I located the bear and began testing the GPS tracking collar over a week ago, but he seems to have left the cave and now I can't find him or the collar anywhere.

In this part . . .

You know some of those commercial programs in Part III that you have to shell out bucks for? In many cases, you can get the same basic functionality (that means without the advanced features found in commercial programs) for free, simply by knowing the right place to point your Web browser. This part discusses how to access and use online street maps, topographic maps, aerial photos, and slick U.S. government-produced maps. You also learn how to save and edit all those wonderful, free Web-hosted maps.

Chapter 17

Saving and Editing Web Maps

• •

• •

*W*hen you need to tell someone how to get somewhere (it could be a party, a house for sale, a company picnic, whatever), Web-hosted mapping services are a great resource for creating maps, especially if you're somewhat artistically challenged. You can get a professional looking map by typing some text and clicking a few buttons.

However, a Web-generated map might not meet your needs exactly. Maybe it's missing a street name, or you want to add a few colored arrows to show a route through a confusing set of roads. Most map Web sites don't have many options to customize a map.

This chapter has everything you need to know about customizing a basic Web map. You'll also find tips and hints on how to make your own maps, using a Web map as the foundation. After you edit the map, you can

✔ Attach it to an e-mail.

✔ Paste it into a word processing document.

✔ Post it on a Web site.

✔ Print it on invitations, fliers, or brochures.

You're not going to become a world-class cartographer overnight, but you can produce some pretty good-looking maps.

The techniques in this chapter apply to any digital map. If you're using a mapping program, you can use the same methods to copy and save parts of any map that's displayed onscreen.

Saving Maps

Most Web-hosted mapping services don't have an option for saving a map to your hard drive. You can print the map, e-mail it, or maybe save a part of it so you can upload to a personal digital assistant (PDA) or cellphone, but there's typically no command to save the map as a graphics file.

Here are three paths around this roadblock:

- ✔ Right-click the map and choose Save Picture As.
- ✔ Use the Print Screen key.
- ✔ Use a screen capture program.

Save Picture As

After you display a map on a Web site, some Web sites let you save the map to disk with a little-known browser command. Follow these steps:

1. **Make the map the size that you want it in your Web browser.**
2. **Right-click the map.**
3. **Choose Save Picture As (or Save Image As, depending on your browser).**
4. **Select the file directory location to save the image.**

   The map picture is saved as a file to the directory location that you select (usually in JPG, GIF, or PNG file format).

Using the Print Screen key

Before Windows appeared on personal computers, the Print Screen key on a PC keyboard sent a copy of the monitor screen to the local printer. When Windows was introduced, things changed. Here's what happens now:

- ✔ **Press Print Screen only.** The entire screen is copied to the Clipboard.
- ✔ **Press Print Screen while you hold down Alt.** The active window is copied to the Clipboard.

Using the Print Screen options are handy if a Web site has disabled the Save Picture As menu command or if the Web map isn't being stored as a graphics file that can be easily saved (such as a series of tiles versus a single image).

Pressing Print Screen or Alt+Print Screen sometimes copies parts of the screen that don't have anything to do with the map, such as menus, title bars, and buttons. If you don't want these on your map, you can either use a screen capture program instead of Print Screen or manually edit the map in a graphics program. This chapter explains both.

Different keyboards label the Print Screen key differently. It might be Print Screen, PrintScreen, PrtSc, or something else, but you should be able to spot it. The key usually is to the right of the F12 function key. (If you're a Mac user, pressing ⌘+Shift+3 saves the screen to a PDF file on the Desktop.)

Follow these steps to save a Web page map with Print Screen:

1. **Display the map in the size you want to save.**

   Map Web sites usually can show maps in several sizes.

2. **Press Alt+Print Screen to save the active window to the Clipboard.**

   This is more efficient than capturing the entire screen with just the Print Screen key. On a Mac, press ⌘+Shift+4, and then press the spacebar. This turns the cursor into a camera. Click the window you want to save.

After the map is on the Clipboard

1. **Run a graphics program such as Paint.**

2. **Use the Paste menu command to insert the Clipboard image into a new file.**

3. **Save the file.**

When a screen capture is stored on the Clipboard, it's saved in bitmap (.bmp) format. Bitmaps don't scale very well, so you end up with jagged lines or a blurry image if you shrink or enlarge the map in your graphics program. Save it as a JPEG or GIF file first. Mac users need to use a graphics conversion program, such as the free GraphicConvertor, to get the PDF file to an editable format.

Using screen capture programs

Screen capture programs provide more options and flexibility than the Print Screen key. For example, you can select part of the screen and copy it. You just save the map without eventually needing to edit all the clutter around it.

Although screen capture programs have different user interfaces, they generally work like this:

1. **Run the screen capture program.**

2. **Assign a *hot key* (usually either a function key or a combination of control keys) that starts the screen capture.**

3. **Make the map the size you want it in your Web browser.**

4. **Press the hot key.**

5. **Click and draw a rectangle over the region of the screen that you want to capture.**

6. **Use the capture command to copy the screen region to the Clipboard.**

   This command will vary depending on the screen capture program you're using. Depending on the screen capture program used, a copy of that piece of the screen is either placed on the Clipboard or directly saved to a file that you name.

Figure 17-1 shows a Web map selected with the MWSnap screen capture program to place in the Clipboard.

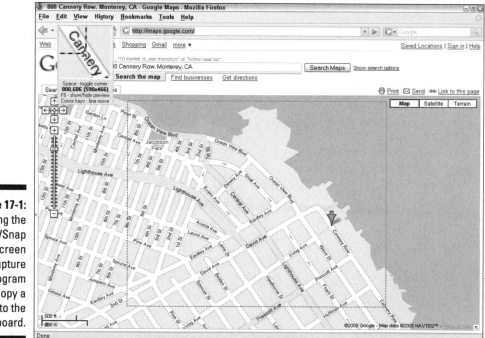

Figure 17-1: Using the MWSnap screen capture program to copy a map to the Clipboard.

Captured: Screen capture programs

Many screen capture programs are available on the Internet. Most are either shareware or freeware. Two of my favorites are

✔ **SnagIt:** A commercial screen capture utility that's been around since the first versions of Windows. It has a number of powerful features. It's available at `www.techsmith.com/products/snagit/default.asp`.

✔ **MWSnap:** If you're on a budget, this is an excellent freeware screen capture utility with lots of features. It's easy on system resources, too. It's available at `www.mirekw.com/winfreeware/mwsnap.html`.

Note: Many screen capture utilities are available for Macs. Do a Google search for "mac screen capture program" to find one that works best with whichever version of the Mac operating system you're running.

Editing a Map

Editing a saved map is just a matter of opening the saved file with a graphics program. The graphics program can be as basic as the Paint accessory program that comes with Windows or as sophisticated as a commercial, high-end design program.

A simple graphics program like Paint is fine if you're not making a lot of changes.

If you have an itch to be a cartographer and want to go beyond using general graphics programs, check out Map Maker, a popular commercial desktop mapping program. You can get a free, limited-feature version called Map Maker *Gratis* at `www.mapmaker.com`.

Almost all Web maps have text at the bottom with copyright information and where the map came from. Leave it intact to respect the copyright and to credit the company that provided the map.

Opening a file

If you used Print Screen or a screen capture program, you need to get the map into the graphics program so that you can edit it. Follow these steps to load a map so you can edit it:

1. **Run the graphics program.**

2. **If one isn't already onscreen, open a new file.**

3. **In the Edit menu, select the Paste command. Or, if you've previously saved the map to a file, use the Open command in the File menu to open it.**

   The map image is loaded, and you're ready to change it.

Cropping

If your map file has more than just your map (such as unrelated windows, icons, or dialog boxes), one of the first things that you need to do after the captured screen is loaded in a graphics program is to *crop* (trim) the image so only the map is shown.

Automatic cropping

Automatic cropping is fast and easy if your graphics program has a crop command. The command automatically removes everything outside of a selected rectangle, leaving only the image within the rectangle intact.

Pixels for free

Although a number of commercial graphics programs are on the market, I'm a big fan of free software. Here are some free graphics programs I find useful for creating my maps:

- **DrawPlus 4:** Simple to use vector program with lots of professional features; www.freeserifsoftware.com/serif/dp/dp4/index.asp.

- **EVE:** A very small, vector graphics program perfect for drawing diagrams and maps; www.goosee.com.

- **GIMP** (Gnu Image Manipulation Program): A popular raster graphics application that's also available for Windows; www.gimp.org.

- **OpenOffice-Draw:** Part of the OpenOffice suite, which means you also get a word processor, spreadsheet, and presentation software in addition to a graphics program; www.openoffice.org.

- **Paint.NET:** A very powerful and friendly Windows raster program; www.getpaint.net.

All these programs are more powerful than Paint, and several offer object-oriented *(vector)* drawing capabilities. Plus, you can't beat the price.

Manual cropping

If your graphics program doesn't have a crop command, you can crop a map manually by copying and pasting what you want. Follow these steps:

1. **Use the selection tool to select the map.**
2. **In the Edit menu, choose the Copy command.**
3. **Open a new file.**
4. **In the Edit menu, choose the Paste command to place the map into the new file.**

Using colors and fonts

The colors and fonts that you use in a map can make a big difference in appearance and readability.

Colors

Colors let a user quickly differentiate features on a map. A black-and-white map might convey the same information as a color map, but hues and shades can make a map easier to read.

Here are some of the most important rules when it comes to dealing with color:

✔ **Use the same general color scheme as the Web map you're editing.**

Any text or features that you add to the map should be the same color as the text and features shown on the Web map.

Use the graphics program's Eyedropper tool to match a color exactly.

✔ **Use contrast between background and text.**

- On a light background, use dark text. Black text is usually best.
- On a dark background, use light text.

✔ **Use bright colors to draw attention to a feature such as a location or route of travel.**

Use red to identify the primary location a user will be searching for.

Fonts

Text labels play an important role in maps; here are some well-founded guidelines for using fonts and typeface styles with maps. These tips will make your digital maps easy on the eyes:

- ✔ **When adding new labels, use the same typeface as Web map text (or as close to the typeface as possible).**

- ✔ **Feature names should be in a normal typeface.** Don't bold or italicize them.

- ✔ **Feature names should only have the first letter capitalized.**

  Capitalizing an entire name on a map makes it harder to read.

- ✔ **Text in a straight line is easier to read and find than curved text (such as a street name that curves around a corner).**

- ✔ **Sans serif fonts are easier to scan on a computer screen.**

  You're reading a serif font right now. Sans serif fonts, such as Arial, have letters without *serifs* (small lines on the top and bottom of some characters).

  If you must use a serif font, Georgia is more legible than Times New Roman.

- ✔ **Don't mix fonts and font size when labeling map features unless you have a really good reason.**

Adding symbols

Symbols convey information without using words. For example, a tent means a campground, a skier indicates a ski trail, and an airplane means an airport.

Most Web maps don't show many symbols. Adding your own symbols is a quick and easy way to improve a map's readability. For example, if you make a map where the route passes by a restaurant, you can add the usual round plate with a knife and fork in the center to show the restaurant on the map.

If you're adding a symbol to a map, make sure that users will understand the meaning. You might think that a picnic table clearly represents a park, but someone else might have no idea what the symbol means and waste a lot of time trying to figure out what it is instead of using the map.

Finding symbols

Here are two ways you can get symbols for a map:

- ✔ **Create your own.** Making your own symbol involves using a graphics program to design a small, icon-sized image.

  If you're not artistically inclined, do-it-yourself icons often result in something that looks like a smudged blob.

✔ **Use a symbol from a font.** Many fonts contain symbols that you can use on your map. Some fonts are composed solely of mapping symbols. Using a font symbol is quick and easy, and you don't need a graphics art degree to make one.

A number of fonts that contain map symbols are freely available on the Internet. The most comprehensive Web resource for finding symbol fonts is at Mapsymbols.com. Check out `www.mapsymbols.com/symbols2.html`.

Commercial GIS software companies offer free programs that can view GIS files created by their full-featured products. Many of these programs install a collection of symbol fonts into the Windows Font folder. (Some demonstration versions of mapping programs do this, too.) Even though you might not use the viewer or demo program, sometimes it's worthwhile installing just to use the fonts.

Inserting symbols

Follow these steps to add a symbol from a font to a map in your graphics program:

1. **Install the symbol font.**

   This process depends on your Windows version. Your Windows online Help system or user guide has instructions for installing fonts.

2. **Run the Character Map program by following one of these two steps:**

   • In Windows XP, choose Start⇨Programs⇨Accessories⇨Syste m Tools⇨Character Map. (In Windows Vista, choose Start⇨All Programs⇨Accessories⇨System Tools⇨Character Map.)

   • Choose Start⇨Run, and then type **charmap** in the Run dialog box.

3. **Select the font you want from the Font drop-down list.**

4. **Click the symbol you want to use.**

5. **Click the Select button; then click the Copy button to copy the symbol to the Clipboard.**

   Figure 17-2 shows the Character Map user interface.

6. **Select the text tool in your graphics program.**

7. **Move the cursor to where you want the symbol to appear and then click.**

8. **Paste the symbol from the Clipboard into the map.**

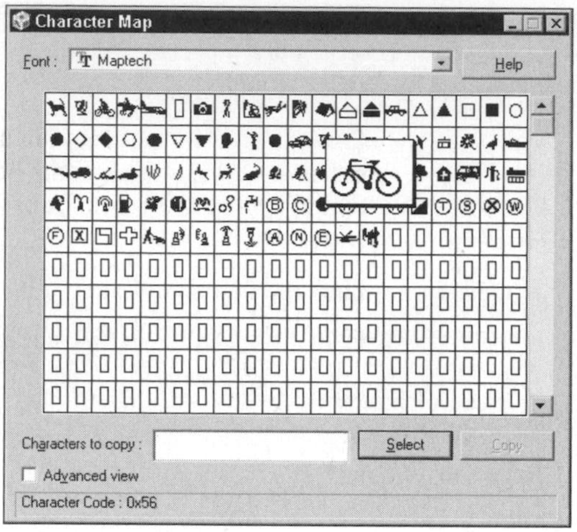

Figure 17-2:
Use Char-
acter Map
to copy a
map sym-
bol to the
Clipboard.

Moving symbols

If you want to move the symbol after you place it, follow these steps:

1. **Click the graphics program's selection tool.**
2. **Select the symbol.**
3. **Drag the symbol to a new location on the map.**

Selecting the right file format

When the map is exactly how you want it, the last step is to save the file. It's important to select the file format that best suits your needs.

If you're sending a map to someone as an e-mail attachment or posting it on a Web page, be considerate of those who will view the map. Keep the file size as small as possible while making the map retain enough detail to be easily read.

The most commonly used graphics file formats are BMP, JPG, GIF, and PNG.

BMP

A *bitmap* (BMP) file is a bit-for-bit representation of how an image appears on a computer monitor. Bitmap files can contain millions of colors.

Don't save your map as a bitmap file if you're using it in e-mail or placing it on a Web site. Files in this format tend to be pretty large because images aren't compressed.

Missing Character Map?

If Character Map doesn't show up in the Windows System Tools menu, choose Start⇨Select Programs and Accessories; some versions of Windows store the command here instead of in System Tools. If you still can't find the program, you need to install it. This is easy and usually doesn't require inserting a CD. Follow these steps to install Character Map:

1. **Choose Start⇨Settings⇨Control Panel.**

2. **Double-click the Add/Remove Programs icon.**

3. **Click the Windows Setup tab and enable either the System Tools or the Accessories option, depending on your Windows version.**

4. **Click the Details button.**

5. **Enable the Character Map option.**

6. **Click Apply.**

Mac users can find a character map in the Font Book utility.

JPG

A *JPG* (or JPEG; pronounced *JAY-peg*) file is designed for saving photographs and other complex images. JPG stands for Joint Photographic Experts Group, the people who came up with this file format.

It supports millions of colors. JPG images are compressed, so they're smaller than BMP files. The JPG format should be used for photographic quality images where most pixels are different colors. You can control the size of JPG images — the higher the quality, the larger the file.

GIF

GIF (pronounced *jif* or *gif)* should be your first file format choice for saving your edited maps.

GIF stands for Graphics Interchange Format. GIF files are compressed but are limited to a palette of 256 colors. This format is widely used for Web maps or any image that has large regions of the same color.

PNG

PNG (Portable Network Graphics) is designed to support millions of colors for Internet use. It's slowly replacing the GIF format. Because it produces relatively small files, you should consider using it for maps that will be sent in e-mail or associated with a Web site.

Designing maps for the Web

Successfully designing maps that are displayed on a Web site isn't just a matter of randomly selecting fonts, colors, and lines and combining them.

If you're serious about creating maps for the Web, check out these three excellent resources:

✔ *Map Content and Design for the Web: A Guide to Optimizing Cartographic Images on the Web.* A Canadian government site with detailed information on Web map design principles.

```
http://atlas.nrcan.gc.ca/site/
       english/learningresources/
       carto_corner/map_content_
       design.html
```

✔ *Web Cartography* edited by Menno-Jan Kraak and Allan Brown. London: Taylor & Francis. A good reference book on how maps are used on the Internet. The book has a companion Web site:

```
http://kartoweb.itc.nl/
       webcartography/webbook/
       index1.htm
```

✔ *Web Style Guide: Basic Design Principles for Creating Web Sites* by Patrick J. Lynch and Sarah Horton. Not map-specific, but important style design guidelines for developing any Web site. There's an online version of this guide:

```
http://info.med.yale.edu/caim/manual/
       contents.html
```

Chapter 18

Navigating Web Road Maps

· ·

· ·

Did you know that free Web services that provide street maps and driving directions consistently generate some of the highest amounts of 'Net traffic? As an example, at the end of 2007, the popular MapQuest Web site produced an average of over 16 million maps and sets of driving directions per day. That's a lot of maps and a lot of people who are using them. And even with the popularity of automotive GPS receivers these days, the number of people visiting map sites continues to grow.

In this chapter, I explore the world of online street maps. Here you discover the basic features that street map Web sites offer, some advantages and disadvantages to using the services, and see the most popular street map Web sites in action.

If you've never used a street map Web site before, this chapter will get you on the road using them. And even if you're a veteran online map user, you'll probably discover some tips as well as other street map sites you might not have used before.

Using Street Map Web Sites

A *street map Web site* is simply an easy-to-use front end to a large database that contains location and map information. When you visit a site (they're all free) and want to view a map associated with a specific street address, the server retrieves the appropriate data from a database and draws a map in your Web browser. You can then scroll and zoom in and out on the map to change views and get other information (such as driving directions) by using your keyboard, mouse, or even a cellphone.

Map sites use *geocoding* to locate an address, which means that street addresses are converted to latitude and longitude coordinates. These coordinates are then passed to the map database to display a map or generate driving directions.

During the first part of 2008, according to market researcher Hitwise (`www.hitwise.com`), the most popular street map sites based on Web traffic were (in order):

- **MapQuest** — `www.mapquest.com`
- **Google Maps** — `http://maps.google.com`
- **Yahoo! Maps** — `http://maps.yahoo.com`
- **Live Search Maps** — `http://maps.live.com`

Listing common street map Web site features

All the popular street map Web sites generally have the same features in common, including

- **Map display:** Street map sites don't display just street maps anymore. Increasingly, the sites are offering a number of different types of maps to help you get around. These include

  - *Street maps:* Streets and roads are displayed as *vector* maps (created with lines and shapes) instead of *raster* maps (scanned versions of paper maps). Vector maps are faster and efficient to display. An example street map is shown in Figure 18-1.

    For places outside North America, map coverage and direction-giving capabilities vary between Web sites. If your favorite map site isn't much help with an international location, try consulting another site.

  - *Overhead views:* In addition to planimetric street maps, Web sites also offer overhead views, using aerial and satellite photographs. Most sites include a *hybrid* view, where streets and street names are overlaid on aerial or satellite photo images. Figure 18-2 shows a hybrid overhead view with street data drawn on a satellite photo.

    Although street maps on Web sites tend to look similar (all the sites use the same map data sources), aerial and satellite photographs can be quite different. Quality and resolution typically vary depending on the location. If I want to get a bird's-eye view of a place, I always visit multiple sites to see which has the best and most usable overhead imagery.

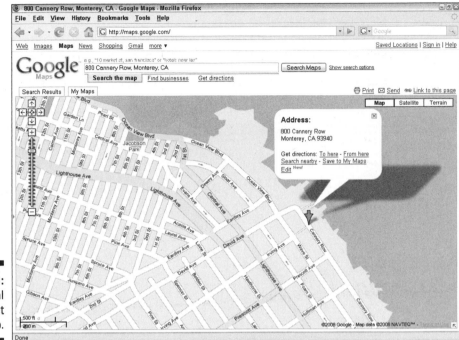

Figure 18-1:
A typical
Web street
map.

Figure 18-2:
A hybrid
street map
with road
and satellite
data.

- *Street views:* Street views aren't really maps but are useful and important to know about. A street view consists of digital photos taken on both sides of a street that is linked to the map data. You can look at a street and see what's really there (shops, sidewalks, planters, and even pedestrians who were caught on camera). Street views only exist for a few urban locations at this point (and not all sites support this feature), but I expect it to be the wave of the future. An example street view is shown in Figure 18-3.

Figure 18-3: With a street view, you see a photograph of the real street.

- *Terrain maps:* Most street maps don't give you a sense of the terrain of an area; is it hilly, flat, or what? To address this, sites have started to use shading to show physical relief on their maps.

✔ **Address searches:** To display a map, you enter a street address, city, state, or ZIP code. The map Web site checks the location information that you enter to see whether it's valid; if it is valid, a map of that place displays.

✔ **Business and service searches:** Street map Web sites also provide you with a souped-up, phone book-style search feature. Enter all or part of a business name and locations that match are shown on the map.

✔ **Route creation:** In addition to displaying a map of a single location, map Web sites can also create routes that show you how to get from Point A to Point B. Just enter a starting and ending address (or city), and the Web site displays a map of how to get to your destination. (Some sites also allow you to create more advanced routes where you can specify intermediate destinations.)

✔ **Route directions:** Street map Web sites also can provide you with driving directions for getting to a location. The directions list streets, turns, and estimated times and distances for traveling between different parts of the route.

✔ **Driving information:** To attract more people to their map services, sites are beginning to provide driving-related information, such as real-time traffic conditions (accidents, construction zones, and average speeds) and current gas prices. An example is shown in Figure 18-4. (Keep in mind traffic information is only available for limited locations; typically in major metropolitan areas.)

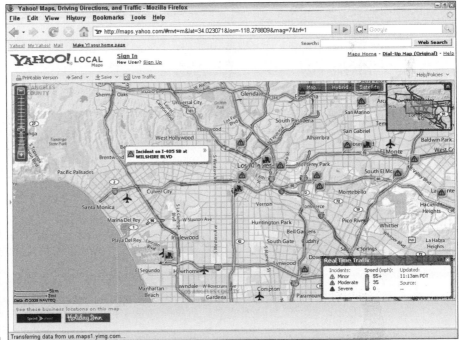

Figure 18-4: Real-time traffic information shown on a map.

✔ **Printing:** All street map Web sites can output maps and driving directions to your printer. (And if you read other parts of this book, you know despite all the cool digital map stuff, I'm still a fan of paper maps.)

Map accuracy

On Christmas Day, 1998, a German driver out for a spin in a GPS-equipped BMW drove his car into a river as he dutifully followed the directions of his in-car navigation system. Despite the stop signs, he continued down a ferry ramp and drove into the Havel River, slowly sinking in about 12 feet of water. Fortunately, the driver and his passenger were not injured. The digital map displayed in his dashboard showed a bridge when it actually should have been marked as a ferry crossing — and the driver trusted his BMW more than his common sense.

Okay, so that was ten years ago, and you'd think GPS and map technology would have improved to prevent incidents like this from happening. Actually, just the opposite is true. It seems like a week doesn't go by without yet another news report of a hapless driver, blindly following his or her GPS unit, who then subsequently crashes into a train, is led on some wild-goose chase down some dirt roads, or suffers some other type of technology-induced misadventure.

These stories illustrate an important point: Street maps will always have some errors, and you can never expect 100 percent accuracy. Software companies, Web sites, and map data vendors do try their best to keep maps current and up-to-date. However, when you consider the 3.95+ million miles of public roads in the United States — with constant, ongoing construction — keeping maps accurate is a daunting challenge. So try not to get too frustrated if a map Web site occasionally makes a boo-boo. More importantly, especially if you use a GPS unit, pay attention and use your common sense!

Web versus PC software street maps

If you've read Chapter 13 (all about PC street-navigation programs), you might be wondering whether you should use a Web-hosted street map or a program that comes bundled with street map data on a CD-ROM or DVD. Both PC and online options can find addresses, display street maps, and create routes. To answer your question, look at the advantages and disadvantages to using street map Web sites compared with their PC program cousins.

Street map Web site advantages

Some of the advantages that street map Web sites offer are that they're

- ✓ **Free:** Street map Web sites are one of those invaluable, free Internet resources. The companies that host these map sites stay in business by selling advertising, offering other mapping products and services to corporate customers, or sometimes charging for premium services.

- ✓ **Easy to use:** Street map Web sites are very simple to use. They offer basic mapping features and easy-to-understand user interfaces.

- ✓ **Available almost anywhere:** If you have an Internet connection and a Web browser, you can find street addresses and create maps from just about anywhere. That even includes Web-enabled cellphones.

✔ **International:** All the sites support mapping outside the United States to some degree or another. PC mapping programs often don't support locations outside North America, or require purchasing several programs to provide coverage.

✔ **More frequently updated:** One of the big selling points to maps on Web sites is they can be updated more frequently than maps that come with PC software, which tend to be updated annually.

Street map Web site disadvantages

Some of the drawbacks to using street map Web sites (in addition to requiring an Internet connection) include

✔ **Limited POIs:** PC street map software packages have a large number of Points of Interest (POIs), such as gas stations, restaurants, and other businesses and services. These POIs, conveniently displayed as small icons on maps, are easy to see and review. Most street map Web sites don't show POIs or only display a limited number, such as selected hotels or other advertisement-based services.

✔ **Limited route capabilities:** Although street map Web sites are starting to improve their routing capabilities by supporting multiple leg routes, PC map programs still have an edge for creating more complicated, multiple destination routes.

✔ **Limited or no trip-planning features:** PC street map programs support calculating trip costs, determining gas stops, setting typical driving speeds, and other advanced trip-planning features. Map Web sites don't have all these options.

✔ **Limited or no customization options:** Unlike PC map software, street map Web sites either don't support or only have simple drawing tools for customizing maps with added text or symbols.

If you want to add text or symbols to a map created with a Web site, you'll need to apply some of the techniques that I describe in Chapter 17 and use a graphics program to edit your map.

✔ **Lack of GPS compatibility:** Unlike a laptop or PDA that's running street-navigation software and is connected to a GPS receiver, Web sites don't provide you with real-time GPS tracking capabilities. (There are some exceptions when using cellphones, which I discuss in Chapter 7.)

✔ **Advertisements:** If you use an online street map, depending on the site, be prepared for lots of ads. Although the ads pay for the mapping service (which allows the service to continue to be free to you), ads can take up a lot of screen real estate, leaving only a small portion available to display the map. Some sites are extremely cluttered with ads compared to others.

Choosing between Web-hosted and PC software street maps

If you can't make up your mind whether to use Web-hosted or PC software street maps, here are few guidelines:

- **Infrequent use:** If you need to look up an address only every now and then and just periodically plan trips, you'll probably be served perfectly well using a street map Web site.

- **Frequent use:** If you travel quite a bit, you'll likely want the additional features that I mention earlier and in Chapter 13 that are found in PC street-navigation programs. These enhanced features make life easier for the frequent traveler.

- **Customizable:** If you want to customize your maps such as adding text labels to them, PC map software offers many more features and options for creating custom maps.

- **Speed:** Even if you have PC map software, street map Web sites can be faster to use if you need to look up a street address quickly, especially if you have a broadband Internet connection.

Reviewing Street Map Web Sites

In this part of the chapter, I list the four most popular street map Web sites, briefly talk about their features, and show what their maps look like.

Companies are always trying to improve the usability of their online maps (as well as get new users by adding new features). You might find some differences in the appearance of the maps that you see displayed on a Web site and the ones shown in this chapter.

To start with, all street map Web sites work the same when it comes to displaying maps and driving directions. In fact, all the sites get their map data from only a handful of providers (such as NAVTEQ and TeleAtlas), so the street maps generally look the same.

Although the map data might be the same, the Web site user interfaces tend to be slightly different. However, if you're familiar with using one street map Web site, you'll be able to easily use others. By mastering the following basic skills, using just about any street map Web site should be a snap:

- **To display a map for a specific location,** you enter an address, city, state (or country), and optional ZIP (or postal) code and then click a search button. See an example of a Google Maps address search page in Figure 18-5.

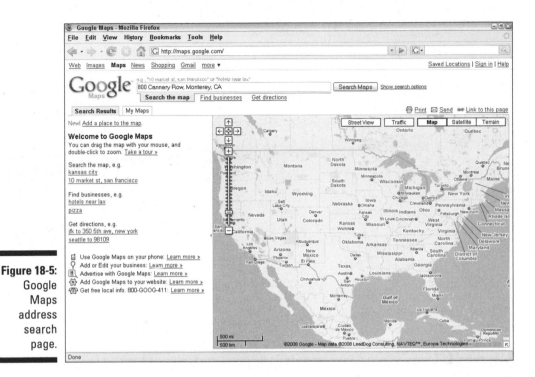

You can also just enter a city and state or a ZIP code to display a general overview map of an area.

✔ **To display a driving route between two points,** either in the form of a map or turn-by-turn directions, you click a directions button or link, and then enter the address, city, state (or country), and optional ZIP (or postal) code for the starting point and destination. As an example, a Google Maps driving directions page is shown in Figure 18-6.

Most sites allow you to hide or shrink the information window that appears to the left of the map. Click the arrow icon between the information window and the map. This enlarges the map and hides the window. Click the arrow again to restore the information window. After a map displays, you can zoom in and out to show more detail or a larger area, respectively. Most maps feature a thermometer-like zoom control with a plus and minus sign at each end. Clicking the thermometer nearer to the plus sign zooms in, and clicking a box closer to the minus sign zooms out.

Depending on the Web site, double-clicking either centers the map on the current cursor location or zooms in on the cursor.

You can move around the map by clicking arrows or compass points that border the map or appear in a compass-like control; the map scrolls in the direction of the arrow or compass point. You can also left-click a map, hold the mouse button down, and move the map by dragging.

At the top of a map, you can change the type of map displayed by clicking the Satellite, Terrain, or Aerial Image button. Available map types vary by Web site.

All map Web sites provide an online help link. Even if you've used a street map Web site for a while, it's worth your time to read the help information. There's a good chance that you'll discover something new that might save you some time.

Take a quick look at the leading street map Web sites.

MapQuest

MapQuest (www.mapquest.com) is the most popular map Web site on the Internet (with Google Maps rapidly gaining on the number one position). The site launched in 1996, and its corporate roots extend back to the 1960s, when its original parent company started making paper road maps that were distributed by gas stations. America Online owns MapQuest.

After you search for an address and display a map, as shown in Figure 18-7, you can print, e-mail, or send the map to a cellphone.

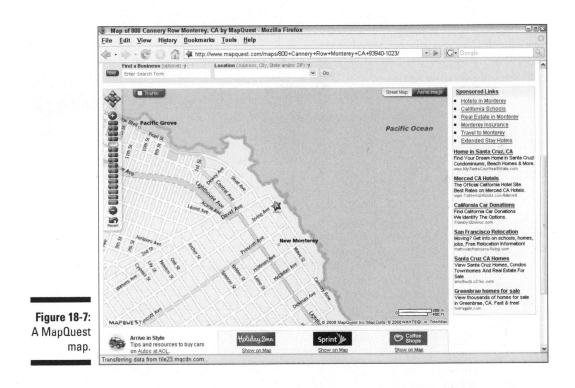

Figure 18-7: A MapQuest map.

MapQuest can provide specially formatted maps and driving directions for wireless PDAs and cellphones with Internet connectivity. Check out the link to Mobile MapQuest (look for Mobile) on the MapQuest home page for information on using different types of wireless devices. (Free and pay services are available.)

Although MapQuest has enjoyed tremendous popularity over the years, other map sites have leaped ahead in terms of features and provide better, less-cluttered user interfaces.

To keep up to date with new MapQuest features, and to learn tips and tricks, visit the MapQuest blog at `http://blog.mapquest.com`.

Google Maps

Google Maps (http://maps.google.com) is a relative newcomer to the online street map world. The site has skyrocketed in popularity by offering a simple, easy-to-use interface and a wide variety of features (and being related to the 'Net's most widely used search engine also doesn't hurt).

In addition to basic street map and direction routing, Google Maps has rolled out a number of innovative features, such as Google Street View (360-degree panoramic street views of major U.S. and Canadian cities), Google Transit (a public transit, trip-planning tool), and Google Ride Finder (an application that shows nearby taxis and limousines in selected areas).

In 2007, Google added a My Maps feature that allows users to create their own maps by drawing lines, shapes, and markers onto a base map. Creating simple, custom maps is easy; and when you're finished, you can save them so other people can view your maps on the Web.

One of the cool things about Google Maps is the relative ease of creating *mash-ups*. A mash-up is 'Net-speak for combining two types of data, usually a map and something else, say apartments for rent, rising sea levels, or crime statistics. (Mash-ups are a little more complex to create compared with the simple maps you can make with My Maps.) Google provides an API (Application Programming Interface) for developers, and the number and types of Google Maps mash-ups available is amazing. The best place to check out existing mash-ups, and keep track of new ones, is Google Maps Mania (http://googlemapsmania.blogspot.com).

If you have a compatible Web-enabled cellphone, you can take Google Maps on the road with you with a number of Google's free, mobile services.

Yahoo! Maps

Street mapping services are a requirement if you're a major Web search engine site. You want to keep users within your domain (both literally and figuratively); you can't send them elsewhere when they want maps and driving directions.

Yahoo! Maps (http://maps.yahoo.com) is Yahoo!'s answer to Google Maps. The site provides street maps (including satellite views), directions, search capabilities, live traffic information, and the ability to send maps and directions to an e-mail address or cellphone. An example map is shown in Figure 18-8.

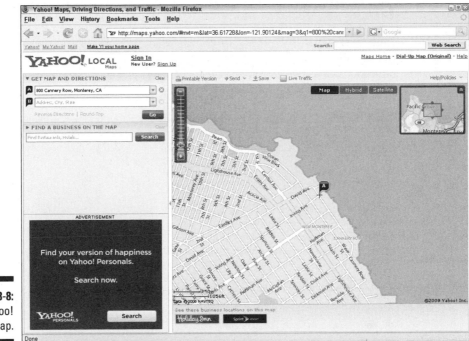

Yahoo! Maps has played follow-the-leader to some of Google Maps' more innovative features over the past several years, but I still feel the site does a good job of offering basic mapping and road information services in an easy-to-use interface with minimal clutter and distractions.

Sergey Chernyshev wrote a nifty Web utility that shows Google Maps and Yahoo! Maps side-by-side, so you can see the differences between the two. Check it out at `www.sergeychernyshev.com/maps.html`.

Yahoo! Maps also offers a developer API and the ability to create mash-ups. Take a look at MapMixer: `http://maps.yahoo.com/mapmixer`.

Live Search Maps

Microsoft's entry in the online mapping sweepstakes is Live Search Maps (`http://maps.live.com`), shown in Figure 18-9. This site has all the features you'd expect (street maps, satellite photos, routing, directions, business searches, and live traffic information). Additionally, the site offers two unique features for certain cities: 3-D views using Microsoft's Virtual Earth technology and high-resolution, color aerial photos taken at an angle called bird's-eye views.

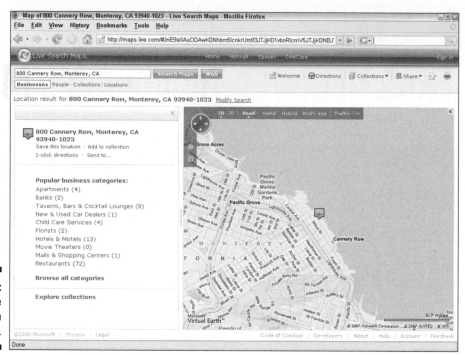

Figure 18-9:
A Live
Search
Maps map.

Chapter 19

Exploring Web Topographic Maps

*W*eb-hosted street maps are handy for finding your way around in cities, but what if your travels take you off-road, on less-beaten paths in forests, mountains, and deserts? You're in luck because a number of Web sites provide free or inexpensive topographic maps of the United States that you can view, print, and take with you on your next adventure.

In this chapter, I discuss some of the differences between topographic maps offered on the Web and dedicated topographic mapping programs that run on your computer. I also present some popular topographic map Web sites and provide basic instructions on using them to display maps.

Using Web-Hosted Topographic Maps

Web-hosted topographic sites are fast and easy to use. If you've already read Chapter 14 (which covers topographic map software for PCs), you might be wondering whether you should buy a mapping program or rely on free or low-cost Web map services. Look at some of the advantages and disadvantages to using topographic map Web sites.

Advantages of topographic Web sites

Topographic map Web sites are pretty handy. Just visit a Web site and *voilà!* — instant map. These Web page maps have several advantages:

- ✔ **Available anywhere:** If you can access a computer with an Internet connection — say at home, the office, a library, or an Internet café — topographic maps are but a few mouse clicks away.

- ✔ **Easy to use:** If you're intimidated by the many options and commands in most PC mapping software products, Web-hosted map sites usually have a basic set of features and commands that are easy to master and use.

- ✔ **Inexpensive:** If you're on a budget, these Web sites are either free or charge nominal subscription fees to access their maps.

Disadvantages of topographic Web sites

Web-based topographic mapping sites have a few limitations:

- ✔ **Small map size:** Most sites display only relatively small-sized (onscreen, that is) maps.

- ✔ **Internet accessibility issues:** Sometimes you might not have access to an Internet connection, the mapping Web site could be down, or your Internet connection could be very slow.

- ✔ **Time investment:** Saving a Web-displayed map (whether you're using screenshots and other techniques described in Chapter 17 or downloading a map from a commercial service) can be a time-consuming process.

- ✔ **No advanced features:** Most topographic map Web sites don't have such advanced features as directly interfacing with a GPS receiver or the ability to change map datums, display terrain three-dimensionally, or customize a map. These features are common in PC mapping programs.

Deciding between Web-based maps and mapping programs

Here are my recommendations as to whether you should use Web-based topographic maps or topographic mapping programs:

- ✔ **Infrequent use:** If you just want to print out a topographic map every now and then, a free Web-hosted map site probably meets your needs.

- ✔ **Frequent or sophisticated use:** If you use topographic maps often, want to create custom maps, or interface your GPS with a digital topographic map, you're better off with one of the map programs described in Chapter 14 (or depending on your needs, perhaps try a commercial Web site that offers topographic maps).

- ✔ **Quickies:** Even if you regularly use PC mapping software, topographic map Web sites are great for quickly creating or viewing maps.

Reviewing Topographic Map Web Sites

Here are some of the Web sites that you can visit to access topographic maps of the United States. I cover both free and commercial sites and discuss some of their basic features. If you're near a computer with an Internet connection, you can follow along as I step through some examples on how to display and use topographic maps on several free Web sites.

Web sites come and go, and user interfaces and site features change, just in case a description and figure don't match what you see on your screen.

Using TerraServer-USA

In 1998, Microsoft, Compaq, the USGS, and several other partners launched the TerraServer Web site. The site was a technology demonstration for Microsoft software and Digital Equipment Corporation hardware, and was dubbed "the world's largest online database." TerraServer had a collection of compressed aerial photos and satellite images that took over a terabyte (TB) of space. (A terabyte, which is 1,024GB, was a lot of disk space back then. Heck, now you can get a terabyte hard drive for your PC for under $300.) Anyone with an Internet connection could freely browse through the database of images.

Now renamed *TerraServer-USA* (www.terraserver-usa.com), this site provides topographic maps and aerial photos supplied by the USGS for just about anywhere in the United States. TerraServer-USA doesn't have a lot of whistles and bells like many mapping programs, but is fast and easy to use.

Most of the aerial images are old, black-and-white photos (although some metropolitan locations have high-resolution *Urban Areas* color images). If you're looking for recent, color overhead photos, visit maps.google.com or get a copy of Google Earth (discussed in Chapter 15).

Displaying TerraServer-USA topographic maps

To display a topo map from TerraServer-USA, you first enter some search criteria. You can search for maps three different ways.

Locations are initially displayed as aerial photographs. To view a topographic map of the area, click the Topo Map tab in the right corner of the window.

Zoom-in searches

When you visit the main page of the TerraServer-USA site at www.terra server-usa.com, you're greeted with a map of the world (as shown in Figure 19-1). Areas displayed in green contain map data. Move the cursor to a location that you want to view and click. Keep clicking and zooming in until the area that you're interested in displays.

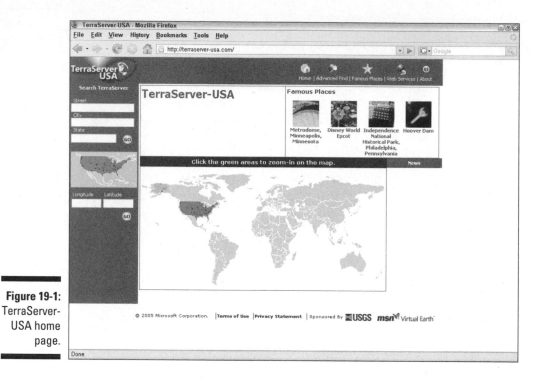

Figure 19-1:
TerraServer-
USA home
page.

Address searches

Enter a street address for the location that you want to view. The City, State, and ZIP Code fields are optional, but using them helps narrow down the search.

Geographic searches

Enter the latitude and longitude coordinates in decimal degrees. As an example, here are the coordinates for Mt. Si (a cool place to hike outside Seattle). Longitude: –121.7401092, Latitude: 47.5076029.

Click the Advanced Find icon at the top of the window to display other search options, including entering the coordinates in degrees, minutes, and seconds.

You can't use TerraServer-USA to search for physical features, such as rivers, mountains, or buttes. If you know the name of a feature, use the Geographic Names Information System to get the latitude and longitude (described in Chapter 12) and use these coordinates for the search.

After you enter your address or coordinates search criteria, click the Go button, and an aerial photo is shown (click the Topo Map tab to display a map). If several images match your selection criteria, you can choose the exact area you're interested in from a list. The topographic map of Mount Si is shown in Figure 19-2.

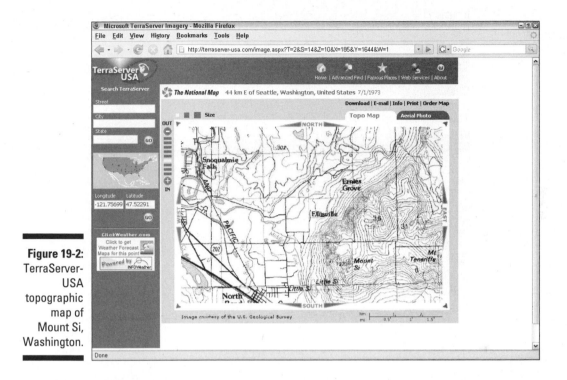

Figure 19-2:
TerraServer-
USA
topographic
map of
Mount Si,
Washington.

Moving in a TerraServer-USA map

When you view a map in TerraServer-USA, a frame with arrows surrounds the image (as shown in Figure 19-2). Click one of the arrows to pan the map in that direction.

Changing map size

You can view three map sizes in TerraServer-USA. In the upper-left corner of the window are three different sized boxes. Click a box to change the size of the displayed map.

Changing map scale

After you display a map, you can change the scale with the zoom control in the upper left of the window. Either

- ✔ **Click one of the bars between the plus and minus buttons.** The bar on the top provides the least amount of detail and largest area, and the bar on the bottom shows the greatest amount of detail and the smallest area.

- ✔ **Click the minus button or the plus button.** This decreases or increases resolution one step at a time, respectively. Figure 19-3 shows the Mount Si topo map zoomed in an additional two levels from the view shown in Figure 19-2.

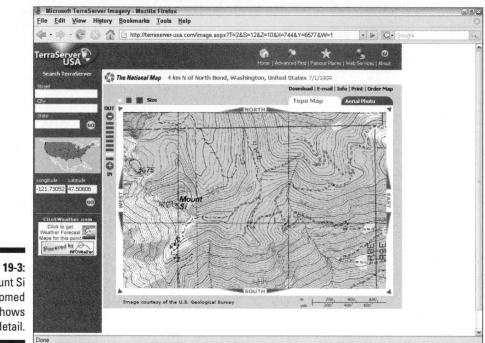

Figure 19-3:
Mount Si
map zoomed
in shows
more detail.

A scale ruler is shown in the bottom right of the window.

You can also use the zoom control to change the resolution of aerial photos. An aerial photo with 1-meter resolution means that you can see an object on the ground that's 1 square meter in size (an example is shown in Figure 19-4). The smaller the resolution number, the more detail the image shows.

Displaying coordinate grids

After you have a map or aerial photo displayed, click the Info link to display latitude and longitude coordinate grids on the map. Clicking Hide Info removes the grids.

Other commands

Other commands above the TerraServer-USA map window include

- ✔ **Download:** Formats the map or aerial photo so you can download it with your browser as a JPG file

- ✔ **E-mail:** Composes an e-mail message with a link to the currently displayed map or aerial photo

- ✔ **Print:** Formats the current map or photo for sending to your printer in either portrait or landscape orientation with a choice of 8.5 x 11-inch or 11 x 17-inch paper

✔ **Order photo:** Jumps to a Web site that allows you to purchase printed copies of the map or photo you have displayed. (Prices vary depending on size and type of image.)

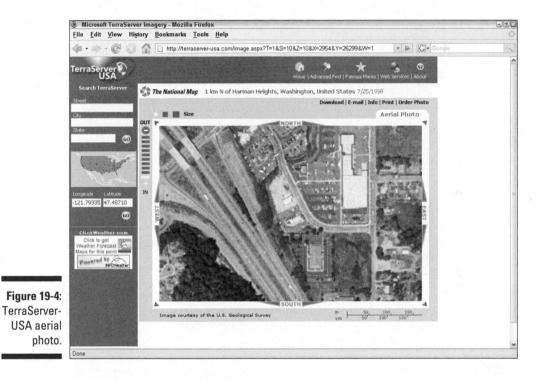

Figure 19-4:
TerraServer-USA aerial photo.

Beyond TerraServer-USA

Although TerraServer-USA is a great, Web-hosted mapping service, it has some limitations:

✔ Only a relatively small geographic area can be displayed at one time.

✔ You can't save a large map area or edit the displayed photos and maps (you can use some of the techniques I discuss in Chapter 17, but if you want a big area, the steps can become both tedious and time consuming).

To address these issues, several developers have written programs that save TerraServer-USA maps and aerial photos to your hard drive as graphics files. These standalone programs have advanced features that allow you to customize and save topographic maps as well as interface with a GPS receiver. One of my favorite free programs of this type is USAPhotoMaps (http://jdmcox.com) shown in Figure 19-5. It's elegantly small and simple to use, with a number of powerful features. If you want to save TerraServer-USA topo maps to print or edit, I highly recommend this program.

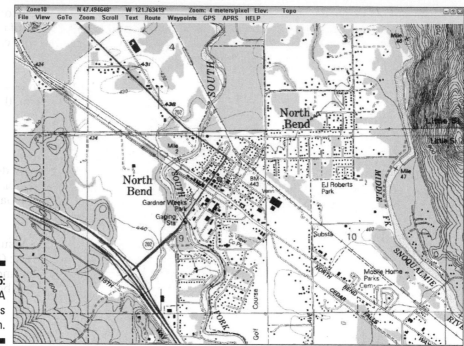

Figure 19-5:
USA
PhotoMaps
screen.

Another online mapping site to try is the Maptech MapServer. The company hosts a free Web service at www.maptech.com to showcase its map technology and sell products. You can find U.S. topographic maps, nautical and aeronautical charts, and search for cities, lakes, mountains, and other features. (As this book goes to press, Maptech is in the process of being sold, and the future of its products and services remains unclear.)

If you're looking for online topographic maps of Canada, check out the Canadian government's Toporama site (part of the Atlas of Canada) at http://atlas.nrcan.gc.ca/site/english/maps/topo/map.

Using GPS Visualizer

GPS Visualizer is a slick, easy-to-use Web-based mapping program written by Adam Schneider that allows you to upload GPS waypoints, track files to a Web site, and then overlay them on topographic and other types of maps.

GPS Visualizer relies on online sources that provide free topographic maps, aerial photos, satellite images, and street maps. Unlike many other sites that are limited to making maps of the United States, GPS Visualizer can create maps of Canada, Europe, and other places in the world.

GPS Visualizer has a number of powerful options for creating customized maps. In this section, I describe how to make a map using some of the Web site's basic features. After you master the general procedure, you can experiment with some of the other advanced options.

If you'd like to follow my examples, point your Web browser to www.gps visualizer.com. GPS Visualizer is free, but if you find it useful, the author requests a small PayPal donation as a way of saying thanks.

Downloading sample GPS Visualizer data

To see how GPS Visualizer works, you need some GPS data. If you don't have any tracks or waypoints handy, head over to www.gpxchange.com (a great site where people exchange waypoints and track files) and download a GPX file for some place that looks interesting.

You can input a number of different GPS data formats into GPS Visualizer, including *GPX* (a standardized format for inputting and outputting GPS data), OziExplorer waypoint and track files, Geocaching.com LOC files, and tab-delimited text files. (Read about OziExplorer in Chapter 16; read about the sport of geocaching in Chapter 8.)

Displaying a GPS Visualizer map

After you have some sample GPS data to use, create a map with it. Enter information in the Get Started Now box (as shown in Figure 19-6).

1. **Click the Browse button next to Upload a GPS File and choose the location of your GPS waypoint or track file.**

2. **Select the type of map to display in the Output Format drop-down list.**

   Options include

   - *Google Maps:* Creates a map using a Google Maps interface.

   - *Google Earth:* Creates a KML format file for use with Google Earth.

   - *JPEG, PNG, SVG:* Saves the map in a graphics file format.

   - *Elevation profile:* Creates an elevation profile graph associated with track data.

   - *GPX file:* Converts and saves the GPS data to a GPX format file.

   - *Text table:* Converts and saves the GPS data to a text file that can be imported into a spreadsheet or database program.

   You can set advanced properties for each of these options by clicking its associated link to the right of the output type drop-down list.

3. **Click the Go! button.**

Seconds later, you have a map! It initially appears as a satellite photo, but you can change it to a topographic map by selecting USGS topo from the drop-down list in the upper corner of the map. (You can select many other types of maps there, too.) The topo map is shown in Figure 19-7.

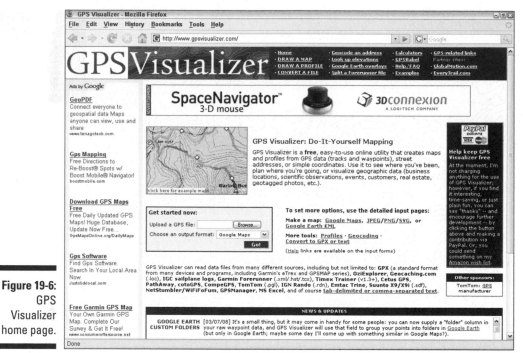

Figure 19-6:
GPS
Visualizer
home page.

The other drop-down list at the top of the map (with percentage numbers) is *opacity*, which fades the background map image in and out.

You can use the Google Maps controls to zoom and move around the map.

That's just the tip of the iceberg for using GPS Visualizer. The Web site has a number of other features and options for making GPS maps, and I encourage you to visit the site and spend some time looking around.

Google Maps currently doesn't display topographic maps, but one topo-related mash-up Web site I really find useful (and fun) is the aptly named Hey, What's That? (www.heywhatsthat.com). Go to a location; the site generates a panorama view showing the hills and mountains you see from that spot, including names, distances, and elevations. (If you have problems with your browser, be sure to read the Web site FAQ.)

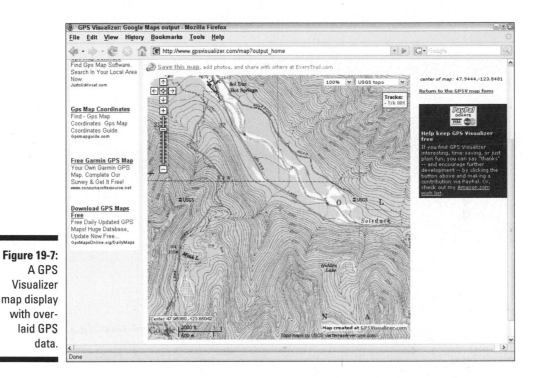

Figure 19-7:
A GPS Visualizer map display with over-laid GPS data.

Commercial topographic Web map sites

Several commercial Web sites sell digital topographic maps and allow potential customers to view some of their maps free (to get a hint of what their for-fee services are like). All these sites tend to have a similar business model, offering

- **Printed maps:** You can buy printed copies of maps you view online.

- **Subscriptions:** On some sites, you can view only certain types of free maps; to access other maps, you pay a subscription fee. Additionally, some sites offer different levels of services. For example, for an extra fee, you might get premium services, such as

  - Downloading maps you can use with PC map programs

  - Creating custom maps

  - Uploading GPS data to use with Web-based maps

The cost of subscription services tends to be comparable or less than the prices of commercial mapping programs that you run on your PC.

When it comes to commercial topographic map Web sites, you're paying for convenience in accessing and using the maps online as well as having a number of different types of maps available in a single place. Although some of the same services are available for free elsewhere on the Internet, commercial map Web sites are priced reasonably enough that you can try them to see whether they meet your needs.

Here are some sites to consider.

TopoZone

TopoZone, the first interactive Web mapping site on the Internet, provides a comprehensive collection of digital topographic maps and aerial photos for the United States. Any Internet users can freely search and view USGS topographic maps. For a yearly subscription fee of under $50, the TopoZone Pro service offers access to aerial photos, street maps, downloading USGS Digital Raster Graphics map, and other enhanced features. To read more about TopoZone (a part of Trails.com), go to `www.trails.com`.

MapCard

MapCard is a commercial Web map site oriented toward outdoor recreation users.

The Web site offers topographic maps, aerial photos, lake depth charts for fishing, and tools for creating custom maps. Although MapCard doesn't directly offer free online maps (a yearly subscription is under $30), the site does have a free "test drive" with which you can download maps and try out the subscription services for a single day. You can visit MapCard at `www.mapcard.com`.

MyTopo

MyTopo (`www.mytopo.com`) is a commercial Web site for creating custom USGS topographic maps or aerial photos. You select a location you want to view (either by place name or coordinates), and MyTopo displays the map or photo. You customize the map by setting its size, selecting a portion of the map to display, and naming the map. A preview copy of the map as it will be printed displays. You can then order a copy of the map in several different sizes on waterproof or glossy paper.

Part V
The Part of Tens

The 5th Wave — By Rich Tennant

So, which is latitude and which is longitude again?

In this part . . .

All *For Dummies* books have a Part of Tens, and this one is no exception. This is a rather eclectic collection of GPS and map-related information. Stuff so intriguing and thought provoking that it deserves a part of its own (or something like that). Anyway, you'll find lots of links (Web sites that is), a primer for printing (maps), and a literal boat-load of marine-related information.

Chapter 20

Ten Great GPS and Map Web Sites

. . . .

In This Chapter

▶ Getting up-to-speed on GPS

▶ Staying current with GPS news

▶ Getting technical GPS information

▶ Downloading free maps

▶ Mapping from your desktop

▶ Modeling terrain

. . . .

The Internet has a tremendous number of GPS and map resources. This chapter contains what I consider the ten best Web sites about GPS and digital mapping. (Actually, there's more than ten, but who's counting?) These Web sites have product reviews, in-depth technical information, friendly people to ask questions, and many sources for free maps to download.

Comprehensive GPS Information

When I'm asked for an Internet GPS resource, I suggest GPSInformation.net (www.gpsinformation.net). Since 1997, Joe Mehaffey, Jack Yeazel, and Dale DePriest have compiled information about GPS, including receivers, software, and antennas. Their comprehensive and frequently updated Web site is packed with technical information, reviews, and common-sense advice for novice to experienced GPS users. This site is a must for anyone who owns or is considering owning a GPS receiver. You'll discover something here, however much experience you have.

Current GPS News and Helpful Advice

The sci.geo.satellite-nav newsgroup is one of the best Internet resources for current GPS-related news. The newsgroup also features a cadre of frequent contributors who provide answers to all types of questions about GPS.

Before you post a question to the newsgroup, such as *What's the best GPS receiver?,* search through old posts with Google for previous discussion threads about your question. GPS newsgroup gurus are generally patient, but they do expect you to do a little homework first. If you can't find any archived information that helps, post a question (and provide as much detailed information as possible). If you're asking which GPS receiver might be the best, provide as much detailed information about your needs as you can.

If you don't have USENET newsgroup reading software or your ISP doesn't carry the `sci.geo.satellite-nav` news feed, you can view and post group messages with a Web browser at Google's cache of this newsgroup:

```
http://groups.google.com/group/sci.geo.satellite-nav/topics
```

Geocaching.com's discussion forums are an excellent source of news and advice. Because GPS receivers are integral to the sport of geocaching (read about this in Chapter 8), check out the GPS Units and Software forum where geocachers share their experience and wisdom with others. You'll find lots of timely news and reviews. Geocachers are some of the most serious civilian users of GPS, so there's excellent real-world information. The forums are located at `forums.groundspeak.com/GC`.

Since the first edition of this book, blogs have become increasingly popular and widespread. I visit several GPS blogs regularly to keep current with new products and information. They include

- `www.navigadget.com`: One trip to the Navigadget site. and you'll be amazed at the number of new GPS products that are released on what seems like a daily basis. Photos, press releases, and commentary provide a "latest and greatest" fix for GPS addicts.

- `http://www.gpsreview.net`: This very comprehensive Web site specializes in reviews of popular handheld and automotive GPS units.

- `http://gpstracklog.typepad.com`: Author Rich Owings' blog has news and commentary along with great product reviews.

- `www.gpslodge.com`: GPS Lodge is another news, commentary, and review blog with one feature I especially like: the weekly "Steals and Deals" posts — links to online retailers offering great deals on GPS units.

A number of GPS receiver user groups are active in Yahoo! Groups and Google Groups that can be excellent sources of information on specific brands and models. To check for a user group associated with your GPS unit, search for the manufacturer's name (Garmin, Magellan, and so on) at `http://groups.yahoo.com` and `http://groups.google.com`.

If you're into boating, I list some marine-specific GPS Internet resources in Chapter 22.

Technical GPS Information

If you're interested in technical, nitty-gritty details of GPS, check out Sam Wormley's GPS Resources Web site. It's a collection of links to sites that get into the science of GPS and other resources that are more suited to the average consumer GPS user. Go to www.edu-observatory.org/gps/gps.html.

Free Maps

Yes, one of the joys of the Internet is that you can download all sorts of free digital maps. Most of the map data has been created by various government agencies and is made freely available to the public. In most cases, you'll need a mapping program to view the maps (described in different chapters of the book), but some of the maps can be opened with graphics programs to print and use. The biggest challenge in downloading free maps is finding them. Here are some Web sites that you should find helpful in your search:

- ✔ http://libremap.org: The Libre Map Project is just what the name sounds like, an effort to make a variety of maps and GIS data available free. The Web site hosts USGS DRG files, 2003 TIGER data, and other maps.

- ✔ www.doylesdartden.com/gis: David Doyle is an environmental geologist (and dart frog aficionado) who maintains an extensive list of U.S. digital map resources. His list at this site is organized by state. You'll find hundreds of links for aerial photographs, topographic maps, geological maps, and many other different types of digital maps.

- ✔ www.macgpspro.com/html/newhtml/maplibrary.html: James Associates makes *MacGPS Pro*, a product for interfacing GPS receivers. The company hosts a list of United States and international free map sources.

- ✔ www.lib.berkeley.edu/EART: Unlike the United States (where digital maps are readily available), in other parts of the world, accurate maps are considered essential to national security and can be difficult to obtain. One of my favorite sources for international maps is the University of California Berkeley's online Earth Sciences & Map Library, where you can download topographic maps for many countries.

- ✔ www.lib.utexas.edu/maps: My other favorite academic map site is the University of Texas at Austin's Perry-Castañeda Library Map Collection. This site has an extensive collection of online maps (as well as links to sites that host maps) for countries and places all over the world.

- ✔ www.un.org/Depts/Cartographic/english/htmain.htm: The United Nations Cartographic Section has an excellent selection of international maps. From doing time-to-time work with humanitarian organizations over the years, another international-flavored site I find useful is ReliefWeb's Map Centre (www.reliefweb.int/rw/rwb.nsf/doc114?OpenForm).

✔ http://oddens.geog.uu.nl/index.php: The most comprehensive collection of map links on the Internet is Odden's Bookmarks. This European Web site has over 20,000 links to maps and map sites all over the world. You can spend hours browsing through links to international map sources.

TIP

An increasing number of states, counties, and cities make their map data freely available over the Web. Do a Google search for a state or municipality's name and *GIS* to check out what types of maps various government agencies are available. A great way to get a return on those tax dollars you contribute.

Expert Desktop Mapping Guidance

Paul Pingrey was a longtime forester who recognized the importance of digital mapping for land management purposes and decided to spread the word. His Digital Grove Web site was a collection of mapping programs, map data sources, and information on how to create 2-D and 3-D maps. Pingrey focused on free and low-cost programs, proving that you don't need to spend thousands of dollars on Geographic Information System (GIS) software or be a GIS professional to produce high-quality maps. It was a sad day when Pingrey announced his site was closing down. The good news is the content lives on at its new home: www.forestpal.com. If you're interested in desktop mapping (and you don't need to be land manager or forester), check it out.

Another frequently updated site that is an excellent resource for both amateur and professional desktop mapmakers is the Free Geography Tools blog at www.freegeographytools.com. Like the name suggests, this site discusses free GIS, GPS, and mapping tools. A number of well-written tutorials are here that walk you through creating some very impressive digital maps.

Finally, if you want to stay current with what's happening in the geo-spatial world (which includes GPS, maps, satellite data, and programs), be sure to visit Slashgeo (http://slashgeo.org).

Definitive Terrain Modeling Information

The most definitive Internet resource on 3-D mapping is John Childs' *Digital Terrain Modeling Journal*. Childs' Web site presents practical how-to tutorials, data sources, and news for creating 3-D maps. Even if you aren't interested in making your own 3-D maps, you should visit this site just to see some of the stunning images that are possible to create by using free and low-cost software. Childs' site is at www.terrainmap.com.

Chapter 21

Ten Map Printing Tips

*I*f you're using a digital map, odds are that eventually you're going to want to print it. Paper maps don't run out of batteries, and they're pretty easy to carry wherever you go. You can just use a program's Print command to transform the bits into ink, but some other techniques make creating a paper map easier.

Make Your Paper Count

The more map and information you can get on a sheet of paper, the better. Most home and business printers are usually fed a diet of 8.5 x 11-inch paper . . . but 8.5 x 14-inch, legal size paper is available, too. One legal size sheet gives you 119 square inches of print area; an 8.5 x 11-inch sheet has only 93.5 square inches of print area.

Some mapping programs can reduce the map scale to get more data on a page. For example, more of the map appears when you print at 75 percent of the original instead of 100 percent. Remember, though, that the detail on the map becomes smaller and more difficult to read when you reduce the scale.

Consider that when you reduce the size of a map that displays a printed scale, such as 1:24,000 or one-inch-equals-a-mile, that the scale will no longer be accurate on the reduced map. That means that plastic rulers and other tools designed for the scale will no longer be accurate.

Print in Color

Maps are usually printed in color because it's easier for our brains to recognize and process colors compared with plain black-and-white. For example, on a United States Geological Survey (USGS) topographic map, you can tell at a glance that lightly shaded green areas are forested and that a solid red line is a major highway. Although navigating with a black-and-white map is certainly possible, using a map that's printed in color is quite a bit easier.

Print the Scale

A map without a scale is like a recipe with only the ingredients listed — not the quantities. You can guess that you add a tablespoon of salt instead of a cup, but the dish tastes a whole lot better if you know the proper amounts. The same is true with a map. If the scale is printed on a map (especially with distance rulers), you can easily judge how far features are from each other. It's also not a bad idea to print information with the datum and declination.

Print UTM Grids

Most mapping software that supports topographic maps can overlay Universal Transverse Mercator (UTM) coordinate system grids on the map. If you're using a GPS receiver with your map, UTM grids make it extremely easy to figure out where you are on the map. Just look at the GPS coordinates of your position, and then count the Northing and Easting tick marks on the map to find your location. (Chapter 2 shows you how to use UTM coordinates to plot locations.)

Use Waterproof Paper

If you plan to use your map outdoors, consider printing it on waterproof paper. Waterproof paper for common inkjet printers uses a vinyl material to produce waterproof, durable, and tear-resistant maps. The paper is a little thicker than regular paper, but you can still fold it. There is also special waterproof paper for copiers and laser printers that serves a similar function.

Waterproof paper costs between 10 and 80 cents per sheet, depending on type and quantity. To prevent wasting expensive paper from printing blunders, first print the map on normal paper to make sure that it looks right.

Never use waterproof inkjet paper in copiers or laser printers unless the paper's package explicitly states that you can.

The heat of their printing processes can melt waterproof inkjet paper and damage the printer.

Here are several sources of waterproof paper for your maps:

- ✔ **iGage:** This company sells waterproof inkjet paper in a number of sizes, including rolls. Its Web site has information on paper specifications and pricing: www.igage.com/WeatherP.htm.

- ✔ **National Geographic:** National Geographic sells waterproof Adventure Paper for inkjet printers. You can order it from this Web site: www.nat geomaps.com/adventure_paper.html.

- ✔ **Ripped Sheets:** This company (www.rippedsheets.com) offers a variety of waterproof paper products, including tags and labels.

- ✔ **Rite in the Rain:** Rite in the Rain is a popular supplier of waterproof field notebooks and paper. If you want to print maps on laser printers or copy machines, Rite in the Rain makes the right kind of paper for you. This Web site has a list of various waterproof paper products: www.riteintherain.com.

You might need to change printer settings to get the best results from using special types of paper. For example, some printers allow you to control ink saturation and darkness, which might produce better images on waterproof paper. And, it might take longer for inkjet printer ink to dry on some special types of paper.

Use a pencil or special pen (such as a Fisher Space Pen) to write on wet, waterproof paper. Regular ink pens don't work too well.

If you're not venturing into the weather, consider using non-glossy, digital photography paper if you have an inkjet printer. The quality and detail is much better than plain paper for topographic maps.

Waterproof Your Plain Paper

If you must use plain-paper maps outdoors, here are some ways to avoid a runny, pulpy mess:

✔ **Carry your maps in a watertight map case or a resealable food storage bag.**

✔ **Create waterproof maps with these options:**

• *Sealers* coat a piece of paper in clear plastic. One of my favorite commercial sealers is *Map Seal*. A 4-ounce bottle coats approximately 8–10 square feet of paper and costs around $7. You can find this at www.aquaseal.com/map-seal.html.

• *Contact paper* is transparent plastic with a sticky back. You can sandwich a piece of paper between two sheets to waterproof the map. Contact paper, which is cheap, comes in rolls so you can cut it to size. Just remember that contact paper doesn't hold up over the long haul.

• *Lamination* involves placing the paper between sheets of a special plastic that seals (when heated) around the paper. Copy centers charge a few dollars to laminate a letter-size piece of paper.

Maps laminated with thick plastic can't be folded. Lighter laminates can be folded.

If you're on a budget, put two maps together back to back and then laminate them. If you're laminating many maps, the savings add up.

Print More Map Area

If you're planning to visit a specific place, add a little more area to your map than you think you need. It's tempting just to print a map that shows exactly where you're going — and nothing more. But what if you decide to take a side trip to some place just off your route that isn't on the map? What if you're interested in a feature you see in the distance? From my years of search and rescue and outdoor recreation experience, I've discovered that you should always have more map than you think you'll need.

If you want a map bigger than what your printer can handle, pay a visit to your local blueprint or copy shop. These businesses, especially blueprint shops, have large-format copier/printers that can make jumbo maps. (Color is expensive, but monochrome is reasonably priced.). Bring a copy of your map on a CD or a USB thumb drive (TIF and PDF file formats are preferred), let the clerk know what size you want it enlarged to, then come back and pick up your poster-sized map when it's done.

You can also make maps smaller. I've used blueprint shops to make reduced size versions of large nautical charts downloaded from the Internet (the official paper charts are a giant 3 to 5 feet long and up to 46 inches wide). Using a graphics program, first convert from color to an easy-to-read grayscale, and then resize the chart to a standard, large-format copier output size (18 x 24 works well). Print the charts you want, optionally laminate or use waterproof paper, and presto! You have your own custom chart book that's much easier to handle on the water.

Put North at the Top

Most maps are orientated with north at the top. When you print a map, follow the same convention. If the map doesn't have a compass rose that indicates directions, you'll know that

- ✔ North is at the top of the map.
- ✔ East is to the right.
- ✔ South is at the bottom.
- ✔ West is to the left.

If you're using a graphics program to edit a digital map, consider adding a north-pointing arrow in case the map doesn't have one. It's a reassuring reminder that north is indeed the direction you think it is.

Use the Best Page Orientation

Before you print your map, decide which page orientation will work best. In your software's Page Setup dialog box, select the orientation.

- ✔ *Portrait* prints the map lengthwise. This is the default for most printers.
- ✔ *Landscape* prints the map widthwise.

Use the Print Preview command (if the software supports it) to see what the printed map will look like. If the orientation doesn't display as much of the map as you want, change the orientation to the other format.

Beware of False Economy

I can remember when gas stations gave away road maps free. (I sometimes catch myself saying, "Back in the days when maps were free . . ." when I'm reminiscing. Boy, does that make me feel old.)

With digital maps, it's easy to think that the days of free gas station maps are back. You can download free maps from the Internet or use inexpensive mapping software, fire up the color printer, and presto! You have a map!

However, don't forget the cost of the paper, ink cartridges, printer wear and tear, and the time involved in printing a map.

Here's an example. An entire USGS 7.5 minute topographic map takes either nine sheets of 8.5 x 11-inch paper or six sheets of 8.5 x 14-inch legal size paper if you're printing at 1:24,000 scale. The costs add up when you're using special paper and lots of color, which empties your ink cartridges. Compare this with shelling out six dollars for an official copy of the USGS map.

Sometimes preprinted paper maps for a few dollars can't be beat.

Chapter 22

Ten GPS Resources for Boaters

"*T*here is nothing — absolutely nothing — half so much worth doing as simply messing about in boats," Ratty said to Mole, in the classic book, *The Wind in the Willows*. I happen to agree, and as a longtime boat mess-around-er, I have to tell you that GPS and boats (of all shapes and sizes) go together like apple pie and ice cream.

In this chapter, I provide a number of GPS and digital charting information resources related to boating. You see how and where to download free nautical charts, get an introduction to marine navigation software, and find out about some great Internet sites that provide lots of details about GPS and other marine electronics.

So welcome aboard. Settle yourself in and all landlubbers are now requested to proceed down the gangplank because we're about to shove off.

For starters, the general convention for reporting marine GPS coordinates is degrees and decimal minutes. This corresponds to chart markings and is preferred by the Coast Guard and professional mariners. Degrees, minutes, and seconds relates to land.

Free Charts to Download

Most newspapers never reported that the U.S. boating world changed dramatically in December 2005. But it did, thanks to the National Oceanic and Atmospheric Administration (NOAA), which made all its digital charts available on the Internet free. Official government paper charts have been around since the 1800s. Digital versions had been available since the 1990s, and although cheaper than collecting a set of paper charts for a large area, they still were expensive. No more. Now, with a computer and an Internet connection, you can download hundreds of current nautical charts free. Just point your browser to http://nauticalcharts.noaa.gov and follow the directions.

Two types of charts are available:

- **Raster Navigation Charts (RNCs):** These scanned versions of paper charts are sometimes referred to as *BSB charts,* the name of the proprietary format back when they weren't free.

- **Electronic Navigational Charts (ENCs):** These are vector charts (see to Chapter 2 for more on the differences between raster and vector). Because electronic navigation is the wave of the future, NOAA is focusing most of its efforts on ENCs. (An example ENC is shown in Figure 22-1.)

A number of free or demo programs are available for viewing and printing charts. NOAA even provides a convenient list of software, including Web links, at http://nauticalcharts.noaa.gov/mcd/Raster/resources.htm.

 NOAA's charts primarily cover coastal waters. If you're interested in major inland rivers in the South and Midwest, the U.S. Army Corps of Engineers has free, downloadable ENC format charts available at www.tec.army.mil/echarts.

Free Windows Navigation Software

After you download a few free charts, you need a program to use them. A number of navigation software packages are on the market that cost anywhere from $50 to $1,000 or more. These programs let you view and print charts, and if you have a laptop computer hooked up to a GPS receiver, you can see where you are on the chart in the comfort of your boat's cabin while cruising.

I'm naturally frugal; therefore, I want to point you in the direction of a great Windows navigation program that's free — SeaClear. (Download it at `www.sping.com/seaclear`.) SeaClear is a powerful and easy-to-use program that works great with NOAA RNC charts. It's in use all over the world and is frequently updated by its creator. A cheap, old laptop (SeaClear isn't resource intensive), free NOAA charts, and an inexpensive GPS receiver give you a bargain-basement navigation system.

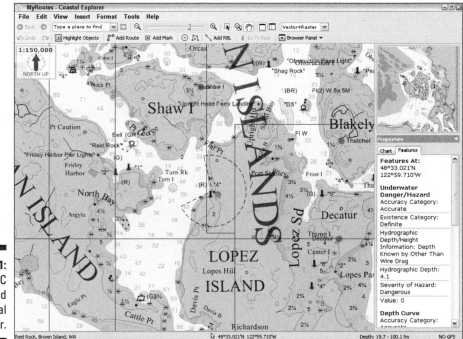

Figure 22-1:
An ENC displayed in Coastal Explorer.

If you want to view NOAA ENC charts, download Caris' free Easy View chart reader (it doesn't have navigation features) at `www.caris.com/products/easy-view`.

Commercial Navigation Software

I like SeaClear a lot. It's a good, basic navigation program and you can't beat the price. That said, a number of commercial software products are on the market that offer more features and functionality than SeaClear (such as interfacing with radar and other devices, displaying tides and currents, and providing weather information). If you're thinking about using a laptop and GPS receiver on your boat, it's worthwhile to check out a few of these navigation programs.

Here's a list of navigation software manufacturers and their products:

- **Rose Point Navigation Systems (`http://rosepointnav.com`):** Founded by former Microsoft employees, Rose Point's Coastal Explorer offers an array of powerful features at a reasonable price. (A Coastal Explorer screen is shown in Figure 22-1.)

- **Maptech (`www.maptech.com`):** Maptech is one of the pioneers in marine navigation systems (before NOAA charts were free, you had to buy them from Maptech). They offer two software packages: The Capn and Chart Navigator Pro (a relabeled version of Rose Point's Coastal Explorer that includes additional chart data on DVD).

- **Nobeltec (`www.nobeltec.com`):** Nobeltec (now called Jeppesen Marine) is a subsidiary of Boeing and produces marine electronics as well as two navigation software products. Visual Navigation Suite is the company's mid-level package and Admiral is their whistles and bells, high-end product.

- **Raymarine (`www.raymarine.com`):** Raymarine also manufactures marine electronics, including their Raytech RNS software for laptops.

- **Furuno (`www.furuno.com`):** Another marine electronics company that produces its own line of software (under the MaxSea label) that interfaces with its electronics products. Very high-end, professional (and costly) software packages.

- **FUGAWI (`www.fugawi.com`):** FUGAWI, which started with land navigation software, offers boaters its Marine ENC and Global Navigator programs.

- **TIKI (`www.tiki-navigator.com`):** TIKI Navigator is a newcomer to the navigation software world but is making waves because of its easy-to-use interface and inexpensive price.

- **GPSNavX (`www.gpsnavx.com`):** If you're a Mac user, you'll want to check out GPSNavX. Affordably priced and highly regarded by boaters who prefer Apple, the program has a lot going for it.

Mad Mariner, a subscription-based, boating information Web site, recently completed an extensive evaluation and review of 15 navigation software packages. If you're in the market for a marine navigation program, this lengthy article is a must read:

```
www.madmariner.com/equipment/electronics/story/HARD_
FACTS_NAVIGATION_SOFTWARE_CONCLUSION_021408_EE
```

Marine GPS and Chart Plotter Manufacturers

Don't feel like risking your laptop to a damp and wet marine environment (especially challenging if you don't have an enclosed boat)? Not to worry, there's a big market for *marine chart plotters* (GPS receivers with large screens that display charts and show your position). These units are specifically designed for use on boats. A chart plotter is shown in Figure 22-2.

Figure 22-2: A Lowrance chart plotter next to a handheld iFinder.

Here's a list of GPS and chart plotter manufacturer Web sites:

- ✔ **Furuno** — www.furuno.com
- ✔ **Garmin** — www.garmin.com
- ✔ **Humminbird** — www.humminbird.com

- ✔ **Lowrance/Eagle** — www.lowrance.com
- ✔ **Northstar** — www.northstarnav.com
- ✔ **Raymarine** — www.raymarine.com
- ✔ **SIMRAD** — www.simrad.com
- ✔ **SI-TEX** — www.si-tex.com
- ✔ **Standard Horizon** — www.standardhorizon.com

Most of the plotters use charts that are preloaded onto memory cards you plug in to the plotter. The cards typically cover a large area. The primary memory card charts companies are Navionics (www.navionics.com), C-MAP (www.c-map.com), and Garmin (www.garmin.com).

If you ever get in trouble and need to radio your position to the Coast Guard for help, be sure it's your actual position. On a regular basis, assistance vessels show up at a set of coordinates only to find no one there. No, the boat in distress didn't sink. It's usually found some distance off after a few radio calls because the captain reported the cursor position on his chart plotter, navigation software, or handheld GPS receiver map. Always double-check that the displayed GPS coordinates are your actual position and not the cursor's.

GPS and Marine Electronics Blogs

Most of the Web sites mentioned in Chapter 20 pertain to using GPS on the land, so if you want to stay current with the latest on marine GPS and other electronics, you need to look elsewhere. I recommend these two blogs:

- ✔ **Panbo:** Magazine writer Ben Ellison has served as editor of this marine electronics blog since 2005, and the site is one of the most popular electronic boat gizmo destinations on the Net. Frequently updated with lots of photos and news on the latest gadgets, it's located at www.panbo.com.
- ✔ **Ask Jack Rabbit:** If you have a GPS unit on your boat, there's a good chance you also have some other marine electronics. Peter James is a marine electrician and the president of Jack Rabbit Marine. He hosts a blog (www.askjackrabbit.com) where readers ask him all manners of questions about boat electronics and electrical systems.

I also mention some marine-related, electronic GPS gadgets in Chapter 7.

Marine Electronics Forums

Blogs are useful, but typically, they only contain one person's opinions and observations. Even if the blogger is knowledgeable, it's still useful to get other views and experiences. Online forums provide a wider range of topics and opinions and compliment blogs, so when it comes to marine electronics, I turn to three popular forums:

- **rec.boat.electronics:** This USENET newsgroup has been around forever and covers all types of marine electronics. You can access the forum through Google Groups at

      http://groups.google.com/group/rec.boats.electronics/topics?lnk=sg

- **Continuous Wave:** A forum devoted to Boston Whalers (as in the boats, not Captain Ahab types), with an informative section that focuses on electronics and electrical topics is at

      http://continuouswave.com/cgi-bin/forumdisplay.cgi?action=topics&forum=
              ContinuousWave:+Small+Boat+Electrical&number=6

- **The Hull Truth:** Kind of catchy Web site name, isn't it? The Hull Truth is a very popular general boating forum that has an excellent marine electronics section. Check it out at

      www.thehulltruth.com

Cruising Wikis

Based on the popularity of Wikipedia (en.wikipedia.org), the online encyclopedia with user-created content, a number of boating-specific wikis have appeared on the Web. These sites allow people to post information on cruising destinations, routes, hazards, and other bits of local knowledge (including GPS coordinates and waypoints). The amount and quality of the data varies, so it's worthwhile to visit the various wikis to see which has the best information for where you typically cruise. Here's a list of wikis and their links:

- **ActiveCaptain** — www.activecaptain.com
- **Captain Wiki** — http://captainwiki.com
- **CruiserLog** — www.cruiserlog.com/wiki
- **MyCruisingLog** — http://mycruisinglog.com

"Proven cruising routes" with compiled lists of waypoints to popular boating destinations are handy, but I recommend creating your own GPS marine routes. Call me paranoid, but I figure lots of people having these widely publicized routes entered into their navigation software or chart plotters just increases the risk of a collision during poor visibility (picture a number of boats of varying speeds all on the same course).

Nautical Chart Overlays for Google Earth

The satellite imagery available from Google Earth is amazing. Even if you don't have an Internet connection, it's possible to preload the satellite images so they're still available, cached on your hard drive. (By using the free GooPs GPS interface program, which I talk about in Chapter 15, with a laptop and Google Earth, you have a powerful, moving map and navigation system.)

Caching images works for coastal and inland marine navigation, but one of Google Earth's slick features is the ability to add overlay data to its satellite images. The folks at EarthNC (`http://earthnc.com`) have done just that, and created nautical overlays based on NOAA charts. They offer two products:

- **EarthNC Raster:** These overlays are based on RNC raster, scanned paper charts. Because you can control the transparency of Google Earth overlays, you can blend satellite photos with nautical charts.

- **EarthNC Plus:** These overlays use all the data on ENC vector charts, such as lights, buoys, and depths. Features overlay on top of Google Earth satellite photos and automatically scale when you zoom in and out.

The EarthNC Web site has more information, including screen captures and sample data.

Keeping Your Handheld GPS Unit off the Bottom

Most handheld GPS receivers are waterproof, and should survive a dunking (the IPX 7 rating, which most handhelds have, means it can be safely submersed for up to 30 minutes in one meter of water). However, being waterproof isn't going to help if you get a case of fumble fingers and your GPS unit takes an accidental swim in deep water.

If you plan to use a handheld GPS receiver on the water very often, especially in a small, open boat, I recommend getting a GPS unit that floats.

Second best is to use a waterproof bag that has some built-in flotation. Voyageur bags (one is shown in Figure 22-3), priced less than $25, are durable, watertight, and buoyant; it's easy to see the screen and use the receiver while it's in the bag. Voyageur, SealLine, and KwikTek all make suitable dry bags. (Search Google to find dealers.)

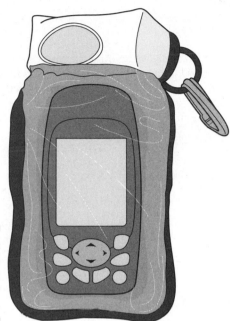

Figure 22-3:
A GPS
receiver in
a Voyageur
dry bag.

My final piece of advice on keeping your handheld GPS unit off the bottom is to use a lanyard that tethers the GPS receiver (or its case or bag) to yourself or the boat. Lanyards are sometimes called "dummy cords," and for good reason. I guarantee you'll feel like a dummy if your lanyard-less GPS unit ever decides to go swimming, and you helplessly watch it disappear into the murky depths.

Although the waterproof Lowrance iFinder Expedition isn't advertised as being able to float, with Lithium AA batteries in it instead of alkaline or NiMH, it actually will. If you have a waterproof-rated GPS receiver you may want to experiment with this idea; I recommend doing so in a sink or bucket for starters.

The battery compartment door is the weak link in any waterproof GPS receiver. I know of several cases where waterproof GPS units were flooded by battery compartments that were improperly closed. If water gets in the battery compartment of a unit that uses memory cards, all the internal circuitry might be exposed to moisture — meaning, bye-bye GPS receiver. Make sure the battery door is sealed, and consider using a waterproof bag as a backup. Also, if your GPS receiver does take a swim, especially in saltwater, be sure to rinse it off with fresh water to prevent possible corrosion.

Free Tide Software

If you boat in coastal waters, knowing the tides and currents can be crucial in getting from Point A to Point B (using GPS, of course). A plethora of printed tide and current tables are on the market, but you should also consider a number of excellent free programs that provide the same information. Here's a list of some of the more popular ones, arranged by operating system.

Windows

- **WXTide** — www.wxtide32.com
- **Wtides** — www.mdr.co.nz

Mac

- **Mr. Tides** — http://web.mac.com/augusth/Tides3/Home.html

Linux

- **XTide** — www.flaterco.com/xtide

Palm

- **Tide Tool** — www.toolworks.com/bilofsky/tidetool

Pocket PC

- **cTide** — http://airtaxi.net/ctide

Index

BUSINESS, CAREERS & PERSONAL FINANCE

Accounting For Dummies, 4th Edition*
978-0-470-24600-9

Bookkeeping Workbook For Dummies†
978-0-470-16983-4

Commodities For Dummies
978-0-470-04928-0

Doing Business in China For Dummies
978-0-470-04929-7

E-Mail Marketing For Dummies
978-0-470-19087-6

Job Interviews For Dummies, 3rd Edition*†
978-0-470-17748-8

Personal Finance Workbook For Dummies*†
978-0-470-09933-9

Real Estate License Exams For Dummies
978-0-7645-7623-2

Six Sigma For Dummies
978-0-7645-6798-8

Small Business Kit For Dummies, 2nd Edition*†
978-0-7645-5984-6

Telephone Sales For Dummies
978-0-470-16836-3

BUSINESS PRODUCTIVITY & MICROSOFT OFFICE

Access 2007 For Dummies
978-0-470-03649-5

Excel 2007 For Dummies
978-0-470-03737-9

Office 2007 For Dummies
978-0-470-00923-9

Outlook 2007 For Dummies
978-0-470-03830-7

PowerPoint 2007 For Dummies
978-0-470-04059-1

Project 2007 For Dummies
978-0-470-03651-8

QuickBooks 2008 For Dummies
978-0-470-18470-7

Quicken 2008 For Dummies
978-0-470-17473-9

Salesforce.com For Dummies, 2nd Edition
978-0-470-04893-1

Word 2007 For Dummies
978-0-470-03658-7

EDUCATION, HISTORY, REFERENCE & TEST PREPARATION

African American History For Dummies
978-0-7645-5469-8

Algebra For Dummies
978-0-7645-5325-7

Algebra Workbook For Dummies
978-0-7645-8467-1

Art History For Dummies
978-0-470-09910-0

ASVAB For Dummies, 2nd Edition
978-0-470-10671-6

British Military History For Dummies
978-0-470-03213-8

Calculus For Dummies
978-0-7645-2498-1

Canadian History For Dummies, 2nd Edition
978-0-470-83656-9

Geometry Workbook For Dummies
978-0-471-79940-5

The SAT I For Dummles, 6th Edition
978-0-7645-7193-0

Series 7 Exam For Dummies
978-0-470-09932-2

World History For Dummies
978-0-7645-5242-7

FOOD, GARDEN, HOBBIES & HOME

Bridge For Dummies, 2nd Edition
978-0-471-92426-5

Coin Collecting For Dummies, 2nd Edition
978-0-470-22275-1

Cooking Basics For Dummies, 3rd Edition
978-0-7645-7206-7

Drawing For Dummies
978-0-7645-5476-6

Etiquette For Dummies, 2nd Edition
978-0-470-10672-3

Gardening Basics For Dummies*†
978-0-470-03749-2

Knitting Patterns For Dummies
978-0-470-04556-5

Living Gluten-Free For Dummies†
978-0-471-77383-2

Painting Do-It-Yourself For Dummies
978-0-470-17533-0

HEALTH, SELF HELP, PARENTING & PETS

Anger Management For Dummies
978-0-470-03715-7

Anxiety & Depression Workbook For Dummies
978-0-7645-9793-0

Dieting For Dummies, 2nd Edition
978-0-7645-4149-0

Dog Training For Dummies, 2nd Edition
978-0-7645-8418-3

Horseback Riding For Dummies
978-0-470-09719-9

Infertility For Dummies†
978-0-470-11518-3

Meditation For Dummies with CD-ROM, 2nd Edition
978-0-471-77774-8

Post-Traumatic Stress Disorder For Dummies
978-0-470-04922-8

Puppies For Dummies, 2nd Edition
978-0-470-03717-1

Thyroid For Dummies, 2nd Edition†
978-0-471-78755-6

Type 1 Diabetes For Dummies*†
978-0-470-17811-9

Separate Canadian edition also available
Separate U.K. edition also available

Available wherever books are sold. For more information or to order direct: U.S. customers visit www.dummies.com or call 1-877-762-2974. U.K. customers visit www.wileyeurope.com or call (0)1243 843291. Canadian customers visit www.wiley.ca or call 1-800-567-4797.

 WILEY

INTERNET & DIGITAL MEDIA

AdWords For Dummies
978-0-470-15252-2

Blogging For Dummies, 2nd Edition
978-0-470-23017-6

Digital Photography All-in-One Desk Reference For Dummies, 3rd Edition
978-0-470-03743-0

Digital Photography For Dummies, 5th Edition
978-0-7645-9802-9

Digital SLR Cameras & Photography For Dummies, 2nd Edition
978-0-470-14927-0

eBay Business All-in-One Desk Reference For Dummies
978-0-7645-8438-1

eBay For Dummies, 5th Edition*
978-0-470-04529-9

eBay Listings That Sell For Dummies
978-0-471-78912-3

Facebook For Dummies
978-0-470-26273-3

The Internet For Dummies, 11th Edition
978-0-470-12174-0

Investing Online For Dummies, 5th Edition
978-0-7645-8456-5

iPod & iTunes For Dummies, 5th Edition
978-0-470-17474-6

MySpace For Dummies
978-0-470-09529-4

Podcasting For Dummies
978-0-471-74898-4

Search Engine Optimization For Dummies, 2nd Edition
978-0-471-97998-2

Second Life For Dummies
978-0-470-18025-9

Starting an eBay Business For Dummies, 3rd Edition†
978-0-470-14924-9

GRAPHICS, DESIGN & WEB DEVELOPMENT

Adobe Creative Suite 3 Design Premium All-in-One Desk Reference For Dummies
978-0-470-11724-8

Adobe Web Suite CS3 All-in-One Desk Reference For Dummies
978-0-470-12099-6

AutoCAD 2008 For Dummies
978-0-470-11650-0

Building a Web Site For Dummies, 3rd Edition
978-0-470-14928-7

Creating Web Pages All-in-One Desk Reference For Dummies, 3rd Edition
978-0-470-09629-1

Creating Web Pages For Dummies, 8th Edition
978-0-470-08030-6

Dreamweaver CS3 For Dummies
978-0-470-11490-2

Flash CS3 For Dummies
978-0-470-12100-9

Google SketchUp For Dummies
978-0-470-13744-4

InDesign CS3 For Dummies
978-0-470-11865-8

Photoshop CS3 All-in-One Desk Reference For Dummies
978-0-470-11195-6

Photoshop CS3 For Dummies
978-0-470-11193-2

Photoshop Elements 5 For Dummies
978-0-470-09810-3

SolidWorks For Dummies
978-0-7645-9555-4

Visio 2007 For Dummies
978-0-470-08983-5

Web Design For Dummies, 2nd Edition
978-0-471-78117-2

Web Sites Do-It-Yourself For Dummies
978-0-470-16903-2

Web Stores Do-It-Yourself For Dummies
978-0-470-17443-2

LANGUAGES, RELIGION & SPIRITUALITY

Arabic For Dummies
978-0-471-77270-5

Chinese For Dummies, Audio Set
978-0-470-12766-7

French For Dummies
978-0-7645-5193-2

German For Dummies
978-0-7645-5195-6

Hebrew For Dummies
978-0-7645-5489-6

Ingles Para Dummies
978-0-7645-5427-8

Italian For Dummies, Audio Set
978-0-470-09586-7

Italian Verbs For Dummies
978-0-471-77389-4

Japanese For Dummies
978-0-7645-5429-2

Latin For Dummies
978-0-7645-5431-5

Portuguese For Dummies
978-0-471-78738-9

Russian For Dummies
978-0-471-78001-4

Spanish Phrases For Dummies
978-0-7645-7204-3

Spanish For Dummies
978-0-7645-5194-9

Spanish For Dummies, Audio Set
978-0-470-09585-0

The Bible For Dummies
978-0-7645-5296-0

Catholicism For Dummies
978-0-7645-5391-2

The Historical Jesus For Dummies
978-0-470-16785-4

Islam For Dummies
978-0-7645-5503-9

Spirituality For Dummies, 2nd Edition
978-0-470-19142-2

NETWORKING AND PROGRAMMING

ASP.NET 3.5 For Dummies
978-0-470-19592-5

C# 2008 For Dummies
978-0-470-19109-5

Hacking For Dummies, 2nd Edition
978-0-470-05235-8

Home Networking For Dummies, 4th Edition
978-0-470-11806-1

Java For Dummies, 4th Edition
978-0-470-08716-9

Microsoft® SQL Server™ 2008 All-in-One Desk Reference For Dummies
978-0-470-17954-3

Networking All-in-One Desk Reference For Dummies, 2nd Edition
978-0-7645-9939-2

Networking For Dummies, 8th Edition
978-0-470-05620-2

SharePoint 2007 For Dummies
978-0-470-09941-4

Wireless Home Networking For Dummies, 2nd Edition
978-0-471-74940-0